Space and Counterspace

A New Science of Gravity, Time and Light

Nick C. Thomas

Floris
Books

Published in 2008 by Floris Books
Fifth printing 2018

© 2008 Nick C. Thomas

British Library CIP Data available
ISBN 978-086315-670-0
Printed by Lightning Source

Contents

*This book is dedicated to my late wife
Heather who so unstintingly gave me her
moral and material support*

1. Introduction

We live in a world full of beauty and wisdom, and to contemplate the beauty of the plants and the behaviour of animals so full of wisdom is surely a source of wonder. The way birds build their nests so securely, for example, with just a beak and claws, is truly remarkable. Not to wonder and marvel at nature is to be poor in soul indeed. Human beings have striven to understand all this as artists, as scientist and as religious people. All three approaches are in their way valid but also one-sided. However great a work of art, it does not explain; however subtle a scientific explanation, it misses the qualitative aspects; however devout a religious approach it remains too 'broad brush,' raising more questions than it answers. But all three approaches are necessary if justice is to be done to our genuinely human experience.

The artist is perhaps least affected by preconceptions, while our current science and religion are full of them. Religion starts from dogmatic premises which are scientifically inaccessible while science blinkers its approach to a narrow physical and material paradigm. The common element to all three lies in the spirit of human beings to strive for understanding. We need to divorce this spirit from dogma and preconceptions and be open, open, open. Why do we have this spirit? Why should our preconceptions put a straightjacket on nature? Nature clearly does not care about them! Why do we let ourselves be channelled into convenient outlooks?

The difficulty science faces is the establishment of objective standards free from arbitrary whim. This is notoriously difficult in the realm of qualities and so science has confined itself to two foundations which are objectively sharable: measurement and number. The successes of science and technology witness the effectiveness of this approach. Galileo first made the distinction between so-called primary and secondary qualities. For example the size and weight of an object are primary qualities while the colour and taste are secondary. By colour here is meant what we directly experience rather than the properties of the electromagnetic waves described by physics. In philosophy the term 'qualia' refers to qualities directly experienced, and a 'quale' is a 'unit of quality' such as a particular shade of blue. Thus Galileo categorized qualia as secondary qualities and any property described quantitatively as primary. The paradox, of course, is that all our science is based on secondary qualities as every observation is based on them, either directly or via instruments. No structure exceeds in reliability the foundation upon which it stands, so the evident lack of understanding of qualia in science could put a question mark over the status of the primary qualities it deduces from them.

Must we for ever give up on the possibility of reaching a scientific approach to qualia? One reason to think not is that the evident success in the use of primary qualities must reflect back on the status of the second- ary qualities from which they are derived. In stripping away qualia, sci- ence deliberately disregards half (or more) of the world of our experience. It is of course free to do so, but at what cost? The alienation many people now feel towards science, the distrust of it, and its manifest inability to handle social problems indicate something of that cost. If indeed qualia can be embraced in a reliable way for knowledge then methods need to be found for overcoming the difficulties. The main problem is that qualia are thought to be subjective, the perceived shade of blue depend- ing upon the person experiencing it, the ambient lighting conditions and the way our senses operate. Difficult yes, but impossible to solve? To say that qualia are all inevitably subjective is an unproved assumption. Indeed the evidence points the other way, for there is much sharing in the realms of music and art, and colours and sounds seem to be sharable. For what is 'blue'? It certainly is not identical with the properties of the electromagnetic waves entering the eye as the same waves can lead to a different experience depending upon the context. Neither is it merely a word, for the word points to something experienced. If it is not to be identified with outer processes, or measurements, or brain processes, all of which are primary qualities, then it must be another kind of entity than any of them. It is more accurate to say that the shade of blue which manifests depends upon the contingencies in force (physiology and sur- roundings) at the time of its appearance. That does not render the shade of blue 'unreal' any more than a potato is 'unreal' because it has grown in a particular kind of soil. If qualia can be objective to some degree, that objectivity refers back to the realm from which they come. There is then 'information' in the fact that one shade manifests rather than another. But the situation seems hopeless for science because it has no inkling of the nature of the realm (or realms) referred to. The subjectivity which bedevils science is not that of the *qualia* themselves but that of the *con-*

tingencies within which they manifest. This is after all well known, and good experiments are designed to standardize those contingencies. But it remains true that the aim is thereby to marginalize qualia.

Particular shades of colour, or particular sounds or tastes or smells are the most elementary cases. Feelings, aesthetic experiences and spiritual experiences are of much greater significance. These concern 'meaning' which is what concerns us as human beings. Alienation, mistrust of science and social incompetence involve this problem of meaning to a large degree. No quantitative extensions of science or technology are likely to solve this for the reasons shown. So it is urgent that the problem be tackled. The work of Max Velmans is very welcome in this regard (Velmans 2000). Both art and religion, unlike current science, seek meaning.

It maybe helpful to explain in some detail the situation vis-à-vis colour. There is an important distinction between our experience of colour and the way colour is handled in physics. Consider the following phenomena:

1. The Moon when eclipsed can exhibit beautiful colours. It may appear to be a rich copper colour, or it may if high in the sky include blues and reds. If it is viewed through powerful binoculars or a telescope those colours may vanish.
2. Observe a red surface in a photographic darkroom illuminated by red light, and it will appear to be almost unsaturated, that is a very light red. It will appear almost white if the colour of the surface is the same as that of the illumination.
3. Observe a dark object against a coloured background, and then remove the object. The background colour where the object was located will appear to be much brighter than the rest.
4. In a darkroom, illuminate a post from two directions, one with red and one with white light. The two resulting shadows of the post are red and green, the green being interpreted conventionally as a kind of illusion produced physiologically.
5. People who are colour blind may confuse colours which are actually distinct.

These and similar phenomena have led scientists to suppose that our perception of colour is subjective, that we do not perceive the 'true' colour, or more radically that our colour experience is a subjective experience of unknown origin outside the domain of science. Physicists study instead the measurable properties of light, which is assumed to be identical to electromagnetic radiation with frequencies lying within a narrow 'window' of the spectrum of possible frequencies.

Thus if I observe a definite shade of blue, there is a *quale* which is the shade of blue I experience, and an associated frequency of the radiation entering my eye. Needless to say qualia are judged to be subjective and the frequency to be objective. The further — often unstated — conclusion is that qualia are not 'real,' being a mere product of ... well, of what? Nowhere in the organism do you find a source of qualia *per se*. There is no known point where nerve impulses or chemical processes get transformed into qualia. How do they arise? What is their ontological status?

We have two quite distinct entities here, qualia and radiation. It is said that potassium red is radiation with a wavelength of 7665 ångströms, that is, three quarters of a millionth of a metre, or a frequency of 391.1 million million cycles per second. So we have two facts: the red qualia and the number 7665. Potassium violet is associated with 4047 ångströms, that is, a shorter wavelength and a correspondingly greater frequency. In each case a quale has been associated with a wavelength of radiation. The latter is objective in the sense that it can be measured and compared by different observers. A parallelism is established between colour and wavelength by a mixture of observation and theory. For example, light is passed through a prism as shown in Figure 1:

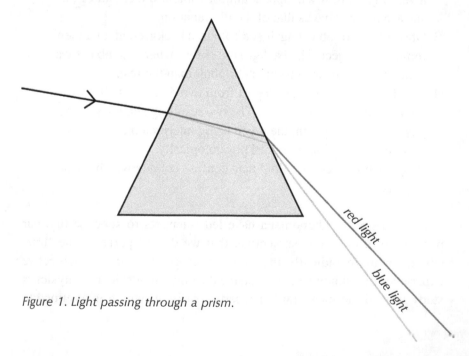

Figure 1. Light passing through a prism.

The total angle of refraction is measured for a given colour. The measurement is repeated for other colours each of which yields a distinct angle. For the same colours the experiment is repeatable. This establishes a parallelism between qualia and angle of refraction. Then theory provides a second parallelism between angle of refraction and wavelength. Combining the two gives that between qualia and wavelength. In practice other methods are used as the dispersion is only about 6 degrees in the arrangement shown, but it illustrates the point being made.

But how do two different experimenters know that they are using the same shade of red? This is solved by noting that incandescent potassium radiates at a sharply defined wavelength which human observers agree is red, any differences in their individual experiences then being irrelevant. Potassium is only used as an example, and other colours emitted by cadmium, potassium, mercury and so on are also used.

The first point to note is that qualia have been eliminated as the last step shows. A robot could in principle be instructed to determine the wavelength of a shade of red by finding it for incandescent potassium. The robot would not need to experience a quale, and 'red' would merely be a label. This proves the point that colour has not been *identified* with wavelengths of radiation, it has only been *put in parallel with them.* All physicists say they know this, but their language often betrays otherwise!

A further problem akin to the phenomena we started out with is that the shade of red of incandescent potassium need not appear the same on all occasions, for instance, if viewed in a room with white walls, or with red ones. The qualia associated with that wavelength are not always the same. If this is true for *one* observer, how are the qualia to be compared for *different* observers? Physics solves this by eliminating qualia altogether, as we have seen. In other words physics does not study colour at all, and only uses the word as a convenience. Some even say red is only a label, as it was for the robot. In any case even the parallelism between colour and wavelength is not exact.

Since our primary experience is of qualia, and corresponding wavelengths are only found by experiencing very complex sequences of qualia (known as experiments), it is clearly illogical to dismiss qualia as 'merely subjective,' or even 'unreal.' They are real enough in everyday life! As they are only approximately parallel to wavelengths of radiation they cannot be dismissed as a mere correlate of wavelength. They clearly exist in their own right in some sense. So how should we assess them?

Heisenberg took this whole issue very seriously and admired Goethe's theory of colour which, unlike Newton's, is a theory of *colour* rather than of radiation, that is, Goethe concerned himself directly with qualia. But Heisenberg says:

> In our advance in the field of exact science we shall, for the time being, have to forgo in many instances a more direct contact with nature such as appeared to Goethe the precondition for any deeper understanding of it. We accept this because we can, in compensation, obtain an understanding of a wide range of inter-relations, seen with complete mathematical clarity.
>
> Goethe, too, sensed an injury in the advance of science. But we may be certain that that final and purest clarity, which is the aim of science, was entirely familiar to Goethe the poet. (Heisenberg 1952, Chap. 5)

In view of all this, we will challenge the idea that qualia are always and necessarily subjective. The very breakdown in the parallelism between them and wavelength suggests they exist in their own right. This is an *empirical* judgment rather than a *linguistic* one. We say this in view of Heisenberg's clearly stated comment in a footnote reflecting a philosophical outlook of his time:

> ... one should particularly remember that the human language permits the construction of sentences which do not involve any consequences and which therefore have no content at all — in spite of the fact that these sentences produce some kind of picture in the imagination; for instance, the statement that besides our world there exists another world, with which any connection is impossible in principle, does not lead to any experimental consequence, but does produce a kind of picture in the mind. Obviously such a statement can neither be proved nor disproved. One should be especially careful in using the words 'reality,' 'actually,' etc., since these words very often lead to statements of the type just mentioned. (Heisenberg 1930, p. 15)

We are reminded of the Vienna school and so-called empiricism in general here. We are, with regard to colour, not concerned with the linguistic use of 'red,' nor of the status of statements about it. We are concerned with our *experience* of it as acknowledged by Heisenberg in connection with Goethe. We particularly cite Heisenberg here in view of his pre-eminence in the field of physics and his clarity of expression.

So what do the senses provide? Briefly we suggest that *their role is to make us conscious of qualia in our environment rather than to 'convey' anything to us*.

This is in a way stunningly obvious if we omit the words 'in our environment.' The other problem raised is the question of consciousness. No consciousness, no experience of colour! Radiation falling on an object need not imply consciousness arises anywhere, for instance, when sunlight ripens fruit. We simply do not know about that and thus leave it aside. That animals respond to colour is well known, and they give every evidence of being conscious at some level, but not necessarily at our own level. We only know in our own case that we must be awake in order to experience colours in our environment. And we also know firsthand that our conscious life consists of the experience of qualia of many kinds as well as those of colour, together with our reasoning ability.

If a man is colour-blind we say from this perspective that his bodily organization is unable to make him conscious of the distinction between, say, green and blue. Similarly for tone deafness.

The degree of consciousness evoked also varies with individuals, for instance, sensitivity to cold. It is also well known that the Inuit experience many more shades of white than British people do, and conversely that we experience many more shades of green. Our respective senses routinely make us conscious of finer distinctions in one realm than in another.

It is further found that what we experience in early life greatly influences what we can experience later on (see, for instance, Jaynes 1990, Chap. 5). *But this is not immutable.* Learning a foreign language as an adult can pose problems because at first we cannot hear certain distinctions between vowels which initially sound the same. But with practice it is possible to learn to distinguish them, which points to the role of thinking in this consciousness adventure. We are told that the distinctions exist and discover we are not conscious of them because we cannot enunciate them correctly, as reported by our teacher. Because we can think, we direct our attention more carefully until we become conscious of the distinctions.

The inseparability of consciousness from the experience of colour (in contrast to a mere reaction to it) supports our thesis, as for other types of qualia. There is now no reason to suppose that qualia are necessarily subjective, for we no longer suppose that they must be conjured up somehow, merely 'parallel' to stimuli. As they are not thus parallel, instead it is clear that we become conscious of them as pre-existing parts or aspects of our environment, as we do of sticks and stones when they break our bones. The 'conjuring' hypothesis is only plausible if there is a strict parallelism. This does not preclude the existence of subjective qualia, of course. We reject the assumption that they are *all* subjective.

Clearly much more could and should be said about this, but enough has been said to make clear how we propose to relate the research described here to qualia, and in particular to colour.

We adopt the following postulate: *We become conscious of qualia according to definite contingencies.*

For example, if I observe a red object in sunlight, I become conscious of a definite quale, whereas if I observe it in a red-illumined darkroom, the different contingencies result in the manifestation of a different quale.

Consider now a definite photon, say emitted by incandescent potassium. The contingency it imposes is very precise so that some shade of red will manifest (apart from colour-blindness). But it is not the only contingency. That of the environment affects it too. We do not say a photon *is* red (or what we will see later is a photon cone). We do say that it imposes very definite contingencies which influence the qualia that can be associated with it. The qualia are themselves distinct. This is perhaps analogous to saying that the contingencies of Earth allow human beings to live on it unaided, but those of the Moon do not. Change the contingencies, for instance, by providing a spacesuit, and the situation changes.

The phenomena cited earlier are all explicable in this way rather than supporting the 'merely subjective' hypothesis. They illustrate the importance of contingencies in determining which qualia manifest. They also show that the qualia in the environment need not all be manifest at any one time: that environment is much richer in qualia than any one set of contingencies exhibit.

The latter comment may indeed indicate something further. Qualia should not be identified with the philosophical concept of 'sense data.' A sense datum is a particular very specific sense experience here and now. But it is defined in the context of the subjective-parallel hypothesis, unlike the view of qualia we are proposing. It may be that qualia are themselves mutable, manifesting according to the contingencies. But that is another story!

To sum up, the work presented in this book will not 'explain' colour but it can throw light on the contingencies required for the manifestation of colour. Similar arguments apply, of course, to sound, taste and so forth.

A spiritual approach to science

None of this is intended to dismiss science or belittle its achievements: what is needed is the *extension* of its remit, and a way of achieving that. Thus a bridge needs to be built from the primary-quality island to the mainland of a fuller grasp of reality. That must begin where it is, namely with the quantitative, and try to establish first an extension of the quantitative to wider realms with the more distant aim of grasping qualia in a sharable manner. This book will not address the qualia problem further, but that problem is nevertheless an essential background to what will be attempted here. The discovery by Rudolf Steiner of another kind of space (Steiner 1920, lecture 15) will be investigated further as quantitative measurements are possible in principle in such a space, and provide a bridge between that space and our everyday one. The distinction between the spaces and their mathematical description depends upon the consciousness of the observer, and consciousness too is, so far, beyond the remit of current science.

Rudolf Steiner (1861–1925) was a spiritual researcher who described the results of that research in 52 written works and nearly six thousand lectures. He claimed to possess extended faculties of perception reaching into other planes of existence that complement the physical plane but are part of one cosmos. Our ordinary waking consciousness, he claimed, gives us access to only a part of the whole, which gives us the possibility of an objective consciousness. As we have seen, the measurable aspects of the cosmos are thereby embraced in a sharable manner, although other qualities are included as well. Our task is to grasp this objectivity thoroughly *and then grow beyond it,* taking with us what we have gained thereby. He experienced the spiritual dimension of the world in a scientific way which he sought to make available to others through his spiritual science. He had a high regard for science, particularly the observational side of science, although he did not agree with its conclusions when based on the materialistic paradigm. He saw the need for modern humanity to base its strivings on a scientific approach, but also that this approach needs to be extended to the spiritual. He gave much advice on how that can be tackled. While many people in previous decades have claimed to have extended consciousness, Steiner is perhaps unique in his strict and scientific approach, and his appreciation of the importance of avoiding the sources of illusion inherent in it. His handling of those sources was comprehensive and full of insight, but also imbued with the appreciation of the strictness necessary to overcome them rather than be cowed by them.

Steiner's particular contribution was to lay a basis for approaching the spiritual in a non-religious scientific manner by examining the very source of our objectivity. For example A.J. Ayer spoke of illusions such as mirages and said that it is the senses which correct them (Ayer 1970, p. 39). Steiner would rejoin that it is because we *think* about the original mirage, and then a later experience, that we correct it. The senses alone make no judgments, only the thinking human being makes judgments. 'That is an oasis' is a judgment, and 'now I see that it was only an illusion' is also a judgment. No such judgments exist without thinking. To grasp thinking as a spiritual activity (Steiner 1992) lays the basis for a non-dogmatic approach to the spiritual, and contrasts profoundly with the view that thinking originates in the brain. Contingency does not prove identity: the fact that I need my brain in order to be conscious and to think does not logically identify thinking with brain activity, just as digestion is not identical with food. Julian Jaynes (Jaynes 1990, p. 39) suggested that we are unconscious of our thinking, and Steiner would agree although he insisted (based on his own experience) that we can wake up to it just as we can waken from a dream to full consciousness. That mysterious activity lying behind thinking can become conscious with training. Now thinking underwrites, as it were, the certainty we have in the use of numbers and measurement, for we are only able to appreciate their power and repeatability because we can think. If we remain stuck there we remain stuck in the current paradigm of science, but if we enter into the thinking activity itself we can use that experience to extend the domain within which science is possible. It must be said that this requires activity beyond mere philosophical criticism, as it must be done just as cooking must be done if a meal is to be produced. Once this awakening has been experienced, philosophical objections to it are as irrelevant as they would be to the process of cooking for a person who has just eaten a good meal.

The work described in this book is heavily 'thinking orientated' and the author is fully aware of that, but the above considerations indicate why that might be justified. It is based on Steiner's insights and cannot be divorced from them, but is not intended in any way as an apologetic. His discoveries need to be worked upon and taken further in a spirit of enquiry, which is what will be described. Just as much scientific work aims to explore Einstein's ideas without thereby being apologetic, so can Steiner's work also be approached. The aim is to build and understand, not justify. A new paradigm is needed to embrace the spiritual aspects of existence, and the work in progress towards establishing such a paradigm will be described in as non-technical a fashion as possible. The mathematical underpinning of this work has already been described in a specialist work (Thomas 1999). It is based on a study of Rudolf Steiner's discovery that spaces can exist which are polar-opposite to our customary space both geometrically and in quality. This book is based on the earlier one, which may be consulted where the reader needs more detailed explanations of what is presented here. Much detailed research has been done since that first book was published, which — it is hoped — will be published later. The point being that what is presented here is neither finished nor static, but part of an ongoing research project. In the next chapter we will describe the other kind of space, which is known as *counterspace*.

Before we do so, a few remarks may be in order on the author's approach to the spiritual. This term means many things to many people, from the religious to the New Age, none of which should be taken to be assumed here. Steiner's approach is based on a kind of inner empiricism, where observation of our thinking activity leads to an individual inner experience of something non-sense-perceptible. If you try to solve a problem and eventually do so, you can review what you did to solve it. You will find that you had to think to do so, and that your thinking activity embraced two essential elements: what we might call 'light' and 'will.' The light gave you an overall grasp of the situation, and the will was expressed in the inner effort you expended. Subtler observation also reveals a 'feeling' aspect. But when you observe this activity of thinking you realize that what you observe has no colour, taste, texture, sound or smell, that is, it is non-sense-perceptible. It is the very first experience of what we will refer to as 'spiritual,' that is, an inner activity depending upon individual initiative. Very much more is possible with suitable training, but this is the starting point. In this book we are primarily concerned with physics and the connection we shall make with the spiritual in this sense concerns the inner effort of will involved. We will speak of stress, which results in force, which in turn relates back to spiritual will in the above sense. We will not be concerned with any schemes or dogma about 'spirits,' although it is obvious that if spiritual activity is possible some entity must be its bearer. We reject the reductionist assertion that thinking is brain activity because that is a theory based on thinking, and no connection has been made between neural activity and qualia or indeed any of the secondary qualities. We know that conscious activity, including perceiving and thinking, require us to use our brains but, as already remarked, that no more means the brain is their generator than the stomach is the generator of the food it processes. It may be more helpful to see the brain as a kind of 'mirror' which reflects into consciousness our thinking activity.

We will also use the term 'ether.' We do not mean by this the nineteenth century concept of a 'luminiferous ether' bearing light and other waves, which later became redundant. Nor do we mean something finer than any known physical substance, for that is a materialistic conception. Following Steiner's research, we have in mind a subtle super-physical realm. However, the development of our ideas on counterspace throws light on the need for modes of reality other than the physical, and indeed explains some of the puzzles in Steiner's descriptions.

2. Counterspace

Rudolf Steiner gave some remarkable scientific indications which are ignored by mainstream science because they do not fit the current paradigm or *modus operandi*. He applied his own extended faculty of perception to aspects of the world hidden from our senses and instruments, and not surprisingly he came up with some interesting and very challenging findings. Those who take his research seriously are put in a strange position by this, for attempts to reconcile them with the findings of conventional research can easily look like a blinkered attempt to hang on to his words at all costs despite other evidence to the contrary. Good science — in theory — proceeds unhampered by such preconceptions. Of course that is an ideal and is not always followed, or there would be no need for paradigm shifts. But does good science also ignore the work of a serious researcher for fear of such criticism? People spend much time and effort in testing Einstein's theories without fear of being regarded as blinkered.

An example may help clarify what is meant here. Steiner discovered with his clairvoyance that light is not electrified in the way science takes for granted, and it does not have a finite velocity but acts instantaneously. The Danish astronomer Ole Roemer (1644–1710) first postulated that light has a finite velocity in order to explain some apparent irregularities in the orbits of the moons of Jupiter. Later experiments seemed to confirm that with ever increasing precision so that we now regard light, and indeed all electromagnetic radiation, as travelling at 299 792 458 metres per second *in vacuo*, this velocity being denoted by the symbol c. The obvious course to follow is to say that Steiner was mistaken. But if we take him seriously we need to see if some way of approaching the contradiction can be found, not to buttress a dogmatic standpoint at all costs, but to see if an important insight of such a researcher may open up other ways of thinking. We need another approach to physics that reconciles the experimental evidence with his findings. The author put this embarrassing state of affairs 'on hold' for many years despite the temptation to take the easy way out by dismissing Steiner's claim. To solve this problem we need to do two things:

1. Explain why light does not have a finite velocity;

   and:

2. Explain why experiments indicate that it does.

As is often the case a 'head on' charge at the problem is not fruitful. Besides, it is not the only problem encountered in Steiner's research reports, so a more comprehensive approach seems necessary than merely remaining stuck on that one point. Steiner also argued against the atomic hypothesis; indicated that the conduction of heat is not as we think; spoke of comets going out of space during their orbits; spoke of lemniscatory paths of the planets instead of ellipses; regarded the Sun as in some sense a region of negative space rather than a massive body; took issue with the Copernican view of the Solar System; spoke about the existence of four kinds of 'ether' in addition to the elements, and espoused Goethe's theory of colour, to name but some examples. Something rather radical is needed to deal with all of this! An apologetic approach is neither fruitful nor scientific.

Instead some lead needs to be followed up objectively that may give new insights. But it must be followed up without merely trying to prove Steiner right on points of detail; the intention rather is to see if a new paradigm will lead to results which throw light on these issues, for or against.

The author chose to follow up work done by George Adams on Steiner's finding that there can be negative space as well as the space we are familiar with (Adams was one of Steiner's collaborators who interpreted his lectures into English and followed up many of his research indications). This seemed so fundamental that it could well be of service. When speaking of the Sun Steiner remarked that if we could visit it by some extraordinary means we would have a big surprise, for we would not find a ball of burning gas (a nuclear fusion reactor in modern terms) but rather a region where there is less than nothing — a negative space. In the nineteenth century the idea was seriously challenged that geometry and space exclusively adhere to the laws set forth by Euclid. Prior to that the Newtonian assumption had seemed the most reasonable, namely that space is an empty and featureless void, while time carries us forward uniformly and is itself similarly featureless. In the nineteenth century two mathematicians, the Hungarian János Bolyai and the Russian Nikolai Lobatchevsky, proved that there can be other types of geometry than the one we learn about at school, and each established a different possibility: either that space is closed or it is 'more open than open.' In other words if we imagine Newton's approach as that of a huge empty box with no sides, continuing indefinitely in all directions, then one different possibility is that space is somehow of finite extent but unbounded, in an analogous way to the surface of a sphere which has no boundaries and yet is closed and finite. This is easier to imagine than the opposite possibility that space opens out more rapidly with distance than the Newtonian box. Euclid's geometry then lies between the two in what amounts to a spectrum of possibilities. So ingrained had been the work of Euclid that these results were not welcomed, and interestingly enough Carl Friedrich Gauss — one of the greatest mathematicians — had in fact discovered this forty years before Bolyai and Lobatchevsky but dared not publish it! However the stage was then set for great advances and Eugenio Beltrami and Felix Klein showed that if there was something wrong with these new non-Euclidean geometries, then there was something wrong with Euclidean geometry also, which forced mathematicians into accepting the new findings.

It might have seemed possible to regard these results as mathematical curiosities of no relevance to the real world, but Einstein based his Theories of Relativity on such ideas and so brought them firmly into the arena of physics. Many television demonstrations, popular books and works of science fiction have made such ideas common coin, although a deep understanding of their true import eludes most people. Thus in Steiner's time a sophisticated and reputable mathematical apparatus existed with which to explore all manner of strange geometries, and George Adams realized that they could be applied to the idea of negative space. This lifted it out of the realm of nebulous concepts into that of mathematical rigour. Of course the establishment of a rigorous geometry is one thing, but showing that it has application to phenomena is quite another. We distinguish between *space* as a factor of the world we live in from *geometry* which describes a space. Whether any actual space corresponds to a particular geometry cannot be settled by argument or speculation, but by empirical experience.

We cannot expect geometry to explain everything, and we need to be clear where we can expect it to lead us. First of all we must distinguish carefully between geometry and physics. Geometry handles forms and spatial relationships, and most importantly transformations, whereas physics is also concerned with force and mass for example. It is hardly to be expected that force and mass are in essence merely geometrical for they evidently have a different quality. Neither can we expect geometry to explain qualities such as colour, sound, taste or other such experiences. Geometry may be expected to define the stage, as it were, upon which these other qualities act.

We shall not expect it to explain colour or any other quality that is non-geometrical, but we will find that it may circumscribe the contingencies for colour to appear. In particular we will find a new approach to light and photons which can only be regarded in this way. We will find that light does not have a velocity, but we will also find out why we think it does, and what the universal constant c really is. We will also find a new theory of gravity which throws light on Steiner's ideas about the Sun.

Negative space is also customarily referred to as *counterspace*. Let us see if we can grasp an idea of it. First imagine yourself standing outdoors on a clear starry night. You can look in all directions, and ideally you occupy one point in the universe with everything else all round you, reaching far away to the infinite distances even beyond the stars. Space seems to extend infinitely far away outwards and in all directions at once. Imagine yourself at the centre of a luminous sphere (say sodium yellow) which is growing outwards from you at an ever increasing rate. As the sphere grows the curvature of its surface decreases, as can be appreciated by comparing the curvature of an iridescent soap bubble with that of the surface of a still lake. The latter has the same radius as the Earth itself, very much larger than the soap bubble. Indeed it appears to be flat, and it is only when we observe the sea horizon carefully that we come to realize it is curved. So as our luminous sphere grows ever larger its surface is getting ever flatter. When only the size of the Earth it already seems locally flat to us. What when it is the size of the orbit of Pluto, or of our galaxy? Very much flatter still! If we allow our sphere to grow indefinitely large, we must conclude that its surface becomes ever more flat until in the limit it becomes quite flat like a plane. But to become quite flat it must *become* a plane, in which case it ends up ceasing to be a sphere at all. This can only happen if opposite points of all its diameters 'meet at infinity' so that it becomes two superimposed planes. This idea is described quite rigorously in projective geometry. We then discard the redundancy of two planes and speak of the *ideal plane at infinity*. This is not part of Euclid's geometry, for in that geometry infinity cannot be reached even in imagination. In the last two centuries the human mind has started to grasp the notion of infinity with some precision, and so in projective geometry we speak of parallel lines meeting in a so-called ideal point at infinity, that is, in a point of our ideal luminous plane at infinity. These points are not like ordinary ones as they cannot be reached physically, and they are added to Euclidean geometry to yield projective geometry. It is as though we closed off our open Euclidean box with a 'lid' which is our luminous plane at infinity. It takes some practice to realize that the lid is really a plane and not a huge sphere. Were it a sphere then it would have an 'outside' and then space would not be closed at all. But a plane has no 'outside' and strangely we find that it is a remarkable plane that only has one side! It must, or space would not be closed off by it in an unbounded way.

So we stand looking at the stars and realize that our ordinary consciousness is poised between a point which is our location in the universe, and a mighty plane beyond the stars and galaxies themselves. That this is so may be regarded as one reason why the non-Euclidean geometries were not welcome. Whether the universe is actually 'flat' and Euclidean, or whether it is open or closed in the non-Euclidean ways we described before, has not been decided by cosmologists. So we may take the liberty of imagining the Euclidean polarity between centre and periphery as we did. This characterizes our consciousness of ordinary space. We now seek to turn this whole picture inside-out to approach a concept of counterspace.

To do this we use the concept of *polarity* frequently used in projective geometry. There is a remarkable symmetry between points and planes, for example three points determine just one plane provided they do not all lie on the same line, for two of them determine a line and then we can imagine a plane turning about that line as axis until it contains the third point. There is only one such plane. On the other hand three planes determine just one point, again provided they do not all contain a line in the way three pages of a book do. An example is the point at the corner where two walls and the floor of a room meet. If we swap the words 'point' and 'plane' in the statement 'three points determine just one plane,' we get the *polar* statement which is also true. Notice that lines play the same role in both cases: the three points must not lie on a line, and the three planes must not share a common line. Conventional science is essentially point-based in its outlook, considering particles which are supposed to be (fuzzy) points, and even reducing fields of force to particle-like discrete entities. Force is supposed to arise by contact, so that if one thing hits another force arises, and fields are reduced to very mysterious particles which hit other particles to cause force. Although the mathematics makes the whole thing appear much more sophisticated, that's what it boils down to. But, as projective geometry shows by means of polarity, every geometrical statement involving points implies another involving planes. Since the point-based approach has been so fruitful in physics we might suspect that bringing in a plane-based one might also be valuable.

This is the basis of our approach to counterspace. Rudolf Steiner experienced it directly through out-of-the-body experiences, but we can approach it by analogy in our ordinary consciousness using polarity. It must be borne in mind throughout that this is only a 'crutch,' but a very useful one.

Polar to the situation of standing at (ideally) a point and looking outwards, we imagine our consciousness rooted instead in a *plane* and looking inwards. When we looked at the stars we arrived at the ideal plane at infinity, so the polar of that will be an ideal *point* towards which we look *inwards*. It seems more natural and certainly accords with experience to say 'inwards.' This ideal point represents an *infinite inwardness* in polar contrast to the infinite outwardness of the plane at infinity. We can sense that the quality of this is quite unlike that of empty space, and inwardness is just what is lacking in our current scientific paradigm. We can now imagine ourselves in the cosmic periphery looking inwards from all directions at once at a yellow spherical surface that is shrinking away from us inwardly ever towards that ideal point without ever reaching it, just as our original expanding yellow sphere could never reach infinity physically. But, we now come to an interesting question about expansion and contraction. In space we can imagine the expanding sphere to increase its radius by one kilometre every second, say, so that we have a series of distances from its centre which mark off equal steps in equal times, and clearly this can proceed as long as we like without the radius ever becoming infinite. In counterspace the shrinking sphere must likewise follow 'equal steps' if we imagine the polar situation, but that is not possible based on radius as we understand it, as equal steps in that sense would soon reduce the radius to zero. That would be a *spatial* interpretation based on our ordinary consciousness rather than a counterspatial one. Instead we imagine some other counterspatial measure which *can* change by 'equal steps' without ever reaching the centre. Now, if we take the reciprocal of the radius we have a quantity that will become infinite when the radius is zero. That is, if we take the radius — say 10 — then the reciprocal is 1 divided by 10. If the radius is one millionth then its reciprocal is 1 000 000 and so on as the radius decreases. 'Equal steps' requires a little bit of maths to explain, but if the radius is repeatedly halved the corresponding counterspace quantity is repeatedly doubled, and the centre cannot be reached in a finite number of steps. This is a spatial model we can think with our ordinary consciousness, but it does accurately model a different kind of quantity in counterspace which we will refer to as *turn*, which measures the separation between planes. Figure 2 illustrates this where the lines shown represent the position of a plane seen edge-on, which is moving through equal steps in counterspace toward its inner infinity at *O*. It will be noticed that the apparent *angle* between successive planes is decreasing, but the turn between them is always the same. However there is no such thing as angle between planes in counterspace, and that appearance arises because we are viewing the thing with our ordinary consciousness. A counterspace consciousness would experience the steps as being equal. Since they are not steps in angle they must be steps for some other measure, which is what we are calling turn. The construction shown in Figure 2 to generate this shows that the plane will never reach *O*, just as a point moving along a line in ordinary space will never reach infinity. The mathematical justification for this was first found by George Adams when he discovered a way of measuring such quantities in counterspace.

In space we also have angles, and there is a counterspace equivalent which we will call *shift* which measures the separation between points. It is a curious quantity which does not appear to have received much attention hitherto. In counterspace we have to be willing systematically to see everything in the opposite sense from our spatial habits. In space two planes meet in a line and we can think of an angle between them, and two points lie on a line and we speak of the distance between them. In counterspace two planes are separated by a turn which is not an angle, while two points are separated by a shift which is not like a distance but in fact behaves just like an angle. We have seen that turn can become infinite, so that if we imagine a red plane tangential to our shrinking yellow sphere then the turn between the red plane and the plane of our counterspatial consciousness increases as the sphere shrinks, until it vanishes into the ideal point, when the turn becomes infinite. On the other hand if we have a blue point *(B)* on a line and another green point

(G) moving along that line then in counterspace the shift between the blue and green points never becomes infinite. Why should it? Generally the line does not contain the ideal point and so infinity does not come into question *in counterspace*. A mathematical analysis shows that if we join the blue and green points *(B* and *G)* to the counterspace infinity *O* with two lines (see Figure 3) then the number measuring the shift σ is the same as that measuring the apparent spatial angle θ between the lines. This is a 'crutch' to help us understand how to visualize the way shift behaves, but remember there are no angles in counterspace. Our ordinary consciousness can use this crutch as long as we remember that we are superimposing on the counterspace situation something that does not strictly belong to it. Although space and counterspace need not be 'back-to-back,' the use of these back-to-back illustrations is helpful *if* we remember that it is just an intellectual crutch. The major challenge for most of us is to be able to grasp these things while remaining in our ordinary consciousness, for the habits associated with it are so ingrained.

Thus we are concerned with something much more fundamental than just geometry, namely the space associated with quite another consciousness. The geometry can give us a description of such a space, but we need to contact in ourselves something that relates to such a consciousness, or shift our awareness while fully awake, to appreciate the qualitative side. To begin by working with polarity is the best bridge for our ordinary consciousness, and we see why Steiner recommended projective geometry as a good starting point.

What is a space?

What characterizes a space? A short answer lies in what it allows and what it forbids as described by its underlying geometry. Certain attributes remain unchanged, and they are of great interest. For example if you drive a car fast you would be most alarmed if it started getting shorter, or thinner, or for that matter bigger. We find that solid objects moving in space retain their volume unchanged, and also the areas of their surfaces, the lengths of their components and the angles between their parts. Furthermore it would be equally alarming if the movement caused the fixed stars to come rushing towards us! For our ordinary experience infinity stays put and geometrical measurements are unchanging or *invariant* during changes such as movement and rotation. How happy some would be if by merely travelling at a different speed their pint of beer was to become two pints! But that does not happen, and the kind of change or *transformation* which leaves lengths, angles and infinity invariant is called a *Euclidean transformation*. It is called that because it is accurately described by Euclid's geometry which we learn at school. This is the geometry of the flat featureless box model of space assumed by Newton. Such a space is called a *metric* space because measurements are possible which relate to what is conserved in the space, the primary quantities in Euclidean space being length and angle. It applies to the movements of rigid solid objects, not to plastic deformations or flow which will occupy us later. When Einstein discovered his Theory of Relativity he made the prediction that objects travelling very fast, say half the 'speed of light,' become shorter, so the space he was working with was not Euclidean.

Counterspace is also a metric space but it is not angle and length that are conserved. As indicated already, turn and shift are its invariant quantities or basic measurements. Why do we refer to the quantity separating planes as *turn*? First consider ordinary space and suppose you are standing on a straight white line and a red ball is rolling along it and away from you. As it does so obviously its distance from you increases until in principle it could end up in the ideal plane at infinity when it would be infinitely far away. Thus we have a line, ideally a point where you yourself are located, and another point moving along the line. If we now imagine the polar situation then we still have a white line, but now you yourself are a plane and there is another red plane which starts off coincident with yours and then rotates 'away' from you about the line. Eventually it will turn so far that it contains (or is contained by, really) the ideal infinitely inward point. We call its separation from you at any instant 'turn' as a picture of this situation. Since the turn eventually becomes infinite we can appreciate that it is not the same as the Euclidean angle between the planes, as angles can only reach 360 degrees.

To travel in counterspace is to undergo transformations which leave the turns between planes invariant. We may wonder about the equivalent of moving in a given direction in space, like a car going along a road. We have to imagine a structure of planes, and the simplest is to take just two planes sharing our white line, say a red and a green plane, with a definite turn between them. As they move about the line that turn stays the same, just as the distance between the front and rear of a car stays the same in space. Figure 2 illustrates this, showing equal steps in counterspace, and also that the corresponding angles are not equal. As we might expect, two planes seem (for our Euclidean consciousness, that is) to get ever closer together as they approach the inward infinity at *O*, analogous in some ways to a tree-lined avenue where the further trees appear to be closer together.

The line passes through *P* at right angles to the diagram and the green and red planes cut the diagram in the green and red lines *(g* and *r)*. If the red plane rotates to the position of the blue line *(b)* and the green plane to where the red plane was, as shown, then the turn between the red and blue planes equals that between the green and red ones. The lines to the left of the vertical axis show how this is constructed practically. The turn is represented by the symbol τ which is the same for all successive pairs of planes as shown.

We must remember that in these illustrations we are looking with our Euclidean consciousness at a point located in our ordinary space and then saying it represents the ideal point at infinity of counterspace. For a counterspace consciousness it does not look like that at all. In fact the tables are turned and in that consciousness, while this point cannot be seen, our ideal plane at infinity can! We will come shortly to the conditions under which our present illustrations are in fact realizable, but for now they are intended as an aid in forming an idea of counterspace.

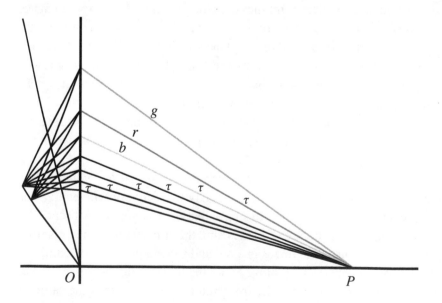

Figure 2. Equal steps in Counterspace.

Counterspace shift is in some ways more difficult to get to grips with conceptually, and yet it turns out it can be handled quite easily. First of all let us think about two parallel planes in space and see what the polar situation is for points in counterspace. Given a red and green plane meeting in a white line, suppose the green plane rotates about some other line which is parallel to the red plane. As it does so the white line moves in the red plane, and when the two planes become parallel the white line moves infinitely far away; it ends up then in the ideal plane at infinity. The polar situation to this last one is a white line containing the ideal point of counterspace, with two points on it, one red and one green. The angle between the two planes has become zero, and likewise in counterspace the shift between the red and green points is zero even though they are distinct. *Two such points in counterspace are 'parallel.'* There are no parallel planes in counterspace, but points can be 'parallel'! Now the very real difference between the two spaces starts to show in earnest. If we have any two points in counterspace, one red and one green, then we may draw two lines joining them to the ideal point, and we have seen that when those lines coincide the points are parallel. It turns out (Thomas 1999, p. 21) that the angle θ between those lines is equal to the shift between the red and green points. Remember, however, that there is no angle *in counterspace* and we are again using our ordinary consciousness to superimpose that angle on the situation. However, that being said, it is a very useful and convenient way of visualizing and handling shift. Finally, we realize that if a third blue point *(B)* is on the line joining the ideal point to the green point, then the shift between the red and blue points equals that between the red and green. Indeed the blue point can move along the line without its shift with respect to the red point changing. This is the polar situation to having a blue plane parallel to the

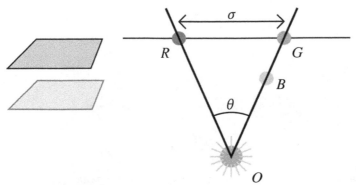

Figure 3. Counterspace shift.

green one (Figure 3), where it is obvious that the angle between the red and green planes equals that between the red and blue.

How, then, is shift conserved in counterspace transformations? Suppose we have a purple line *(p)* in counterspace and two points on it, let's say one red and one green again *(R and G),* and we draw the lines joining them to the ideal point, say a blue line *(b)* through the red point *(R)* and a yellow one *(y)* through the green *(G).* If a transformation moves the purple line to some other position *(p')* then one way for the shift to be conserved is for the red and green points *(R and G)* to move along the blue and yellow lines *(b and y)* respectively to the new inter-sections of the purple line *(p)* with the blue and yellow lines *(R' and G').* Another way is for them to move to new positions such that the angle between the resulting new blue and yellow lines equals that between the old, as in Figure 4.

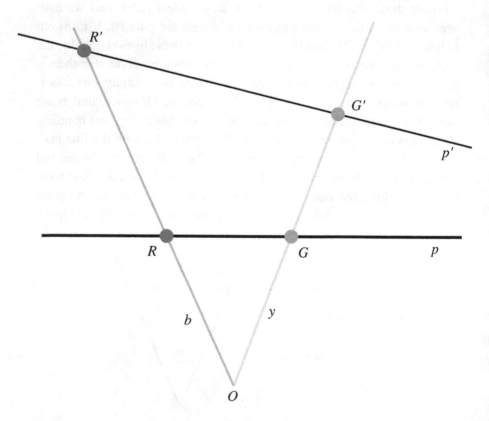

Figure 4. Conserving shift in counterspace transformations.

A full counterspace transformation is such that all turns and shifts remain unaltered just as all lengths and angles do in space. However, in space volumes and areas are also unaltered. Can we find their equivalents in counterspace? First we must appreciate that the roles of 'inside' and 'outside' are reversed. The surface of a sphere partitions ordinary space into a finite volume we call the 'inside' and the rest of space which we call the 'outside.' Notice that the outside contains the ideal plane at infinity. Then in counterspace we expect the 'outside' to contain the ideal point, so that what ordinarily looks like the inside is now the outside and vice versa. Thus for counterspace what ordinarily looks like the rest of space is a finite region, and in fact makes up its *polar-volume* as we shall call it. Furthermore while in space a sphere has a centre which is a point, in counterspace it has a polar-centre which is a plane. If its ordinary centre coincides with the ideal point of counterspace then its polar-centre coincides with our plane at infinity. A similar argument applies to the inside and outside of a cube or a tetrahedron or any other closed surface. If we have an ordinary cube with sides of length s then we calculate its volume as $s \times s \times s$. In counterspace we carry out a similar calculation, but using turn instead of length, so that if the turn between two opposite faces is T then its polar-volume is $T \times T \times T$ which is thus a measure of that region of counterspace not containing the ideal point. These calculations are shown in detail elsewhere (Thomas 1999, Chap. 4), and we simply present a plausibility argument here. The situation is really quite conveniently simple: if we have a formula for the volume of an ordinary geometrical object in space, then the same formula applies to its polar form in counterspace but using turns and shifts instead of lengths and angles. We should really expect this if we have done our polarizing properly. This also throws light on the name 'polar Euclidean space.'

Similar arguments apply to *polar-area* in counterspace, the formulae again being the same as for space. We will take one example which proves to be important later, and that is the polar form of a circle. Let us build this up carefully. A circle lies in a plane and consists of all points at the same distance from a fixed point (its centre). Its polar will 'lie in a point' and thus consist of all planes containing that point which have the same turn from a fixed plane (its polar-centre). The result is a circular cone, not like a dunce's cap but rather the complete cone (see Figure 5). If the ideal point of counterspace lies on the axis then the polar-area is the region of counterspace not containing it, and the polar-centre is the plane in its apex at right angles to the axis. Again we find a challenge to

our normal intuition as that region looks like an infinitely large volume!
It is not infinite for counterspace, though, and is not a polar-*volume* but
a polar-*area*, which can be calculated. The reason it is not a polar-volume is that the planes making up the polar-area are not fully free but are
constrained always to contain a fixed point (polar to the points inside
the circle all lying in a fixed plane). The tangent planes to the cone cor-
respond to the points on the circumference of the circle, and the turn
between the centre-plane and each of those tangent planes is the same
for all of them and is the *polar-radius* of the cone.

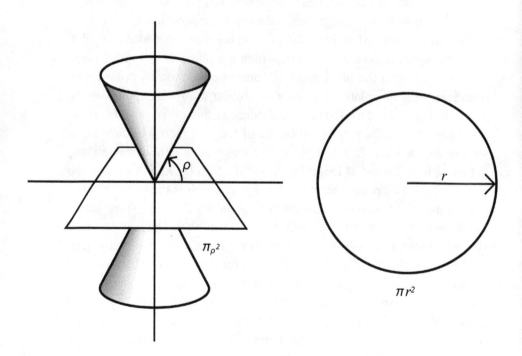

Figure 5. Polar of a circle and area of a circle.

It is now clear that transformations which leave turn and shift
unchanged will also leave polar-area and polar-volume unchanged. If
there is any doubt about this, note that we are using the same formulae as
for space, and we know those formulae give unchanged results in space
since volume and area are conserved there, so it follows they must do
so in counterspace as the corresponding turns and shifts are unchanged
by a transformation. Remember, so far we have concerned ourselves

with the way rigid solids such as crystals may behave in space and their equivalents in counterspace. We have not yet concerned ourselves with liquids or gases, which are discussed in Chapter 6.

One point to stress before we go on: we do not envisage counterspace as being in any way 'back-to-back' with space. It is a completely distinct space with its own laws and *need* have no interaction with ordinary space at all ('back-to-back' means that the two spaces would effectively coincide for all points, lines and planes). Of course a particular counterspace *could* exist which fitted back-to-back with space so that all planes of space were also planes of the counterspace, *but that need not be the case* and we assume it is not. The back-to-back assumption easily creeps in because we use our ordinary spatial imagination as a crutch to visualize counterspace. However, there must be some relationship between the spaces.

We envisage counterspace as a real factor in the world, in other words we assume the actual existence of one or more counterspaces just as we assume the existence of the space we are used to. However, that does not necessarily mean that these spaces 'know' about each other, or that there is an automatic or inherent relationship between them. Such a relationship has to be created in some way so that what happens in one space influences what happens in the other, as though 'doors' or 'windows' are established between the two, or perhaps 'stitches'! As an analogy, suppose that you have a map of your country. If you puncture or obliterate a city on the map then, unless you are some kind of magician, that will not have any effect on the real city. The reverse aspect of this truth can be rather annoying, where something has changed in reality since the map was made, such as a road being blocked, and you do not find what you expected and suffer inconvenience as a result. Were the map linked to the country in some real way, then changes of one would affect the other. Dowsers claim (perhaps by implication) to effect such a linkage when they use maps to find lost or hidden objects.

3. The Relation Between Space and Counterspace

We are saying that space and counterspace are initially just as unrelated as the map and the country; indeed we go further as we do not suppose that there is any formal similarity as is intended with the map. But if counterspace is a real factor of our world and not just a mathematical fiction then it can only have significance if there is some kind of traffic between it and space. We thus propose that:

space and counterspace are related by definite kinds of linkage.

What is interesting is the nature of these linkages and how they are made and sustained. This work started out as the exploration of an idea, as we said earlier, and the first part of that idea is that space and counterspace are linked by sharing either common points or common planes, although we will see later that for the ethers something more subtle is needed. If the spaces are linked at points then this is rather like a 'stitching' of the two together. It need not mean that *all* their points are linked, or all their planes. They may just be linked at special points, so that what happens at other non-linked points is the affair of only one space, depending in which one it exists. The lining of a coat is usually only attached at special points, and is independent of the coat elsewhere. When we say the spaces are linked, then, we mean that anything in space that tries to move that point will have a corresponding influence on counterspace as its corresponding point will be affected, and similarly for linked planes.

Things get more interesting when we have a number of linked points forming some kind of shape, as then we start to wonder what will happen if the whole shape either moves or is distorted. The reason why this might be interesting is that the laws of counterspace are quite different from those of space, so what may be a lawful change of the form in one space need not be so in the other. Let us consider a cube and suppose that its eight corners and six faces are linked. Now what characterizes our ordinary space is that solid objects do not change their shape or size when they are moved about. As we remarked before, you would

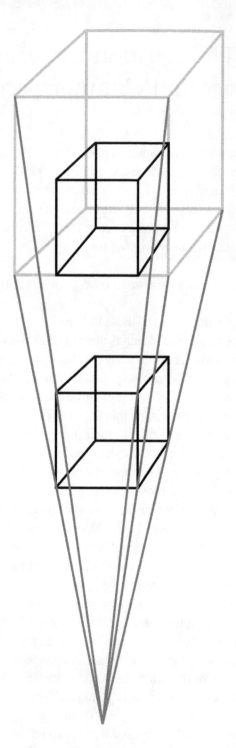

Figure 6, When moving a cube in counterspace it expands.

find it disconcerting if your car became shorter or thinner when you drove it fast on a motorway! We do not expect cars to behave like that. This property is called *invariance* by mathematicians, and specifically our ordinary space is characterized by those properties of solid objects that remain invariant. These are length, angle, area and volume as we expected for the car. Thanks to this rulers and protractors are useful practically, or great confusion would result if we could buy a longer piece of cloth for the same price by measuring it with our ruler vertical rather than horizontal! This concerns rigid solids, of course, and does not apply to a balloon, for example, which gets bigger as it rises through the atmosphere. Other considerations must then be included.

The same applies to the linked cube. So if we move it upwards we expect it to remain the same size and shape. But what about its behaviour in counterspace? If we suppose that the infinity of counterspace is centrally located underneath it, then as it moves away from that very special point (the ideal point) it will try to respect the invariants of counterspace. We saw in a previous chapter that in that space shift and turn play the primary roles, and they are the invariants for counterspace. Thus the shifts between the corners of the cube will try to stay the same, and as illustrated in Figure 6 that requires the cube to expand as seen in space.

This is because the shift between two points on each of two lines through the counterspace infinitude O remains the same if those points stay on those lines. Thus the corners of the cube must do this, and that requires the cube to expand in ordinary space.

Thus if we move the cube we end up with a problem in one of the spaces, or even in both. The result is a geometrical distortion of the cube, for instance, if it remains faithful to counterspace. We will refer to this as *strain*, a term borrowed from engineering, so that in space the cube suffers strain as a result of the movement. Had it remained invariant in space the strain would have been in counterspace. This is the first aspect of our central idea, that:

objects linked to both space and counterspace may suffer strain as a result of transformations.

The second part of our idea concerns what results from this strain. To see what we have in mind, consider a metal rod stretched by a force (Figure 7).

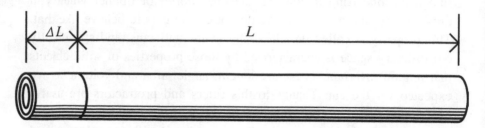

Figure 7. Stretching a metal rod.

Suppose we have a metal rod of length L which we subject to a force which then stretches it so that the extension in length is ΔL. We say that the strain is $\Delta L/L$, which is a purely geometric definition of the proportional change in length. But the rod responds by resisting what we are doing, and internally there arises what is technically called *stress* existing throughout the volume of the rod; it results in a force which balances the applied force. This is not a geometrical quantity, and it depends upon what the rod is made of because a greater force is required to stretch steel than copper, assuming both initially have exactly the same cross sectional area and length. Thus the two rods of different metals may suffer the same *strain* but different *stress*.

We now adopt this distinction for linkages between space and counterspace, so that the second part of our central idea is that when the linked cube suffers strain as a result of a transformation then there is an answering stress, so that we say:

A transformation of an object linked to both space and counterspace may result in strain which will call forth an answering stress.

The difference between strain and stress is of very great importance, because while strain is a purely geometric idea, stress relates to force which involves new non-geometric factors. *We make the transition from geometry to physics when we go from strain to stress.* We are not going to suggest that everything can be described solely by geometry, but rather we recognize that quite other factors come into play when stress arises.

Thus we have a threefold situation, for we have space and counterspace and then arising from their linkage stress emerges as a third factor. At this stage we leave open any questions about the origin of the stress.

This opens up the possibility of including in science, in a rigorous and objective manner, the action of non-physical agencies as part of our world. Thus if there exist forces whose origin is difficult to account for physically it may be that it lies in counterspace due to stress there. In our investigation of these ideas, the first such possibility examined in this light was gravity and, as we shall see in Chapter 5, it is possible to account for it in terms of a stress arising from the rate of change of a counterspace strain. This finding encouraged further work exploring the consequences of our central idea, which proved more fruitful than anticipated.

Much of modern physics is difficult to understand physically, and it has tended to become mathematical in nature, making no attempt to explain what happens, but restricting itself to predicting the results of experiments and interactions. It declines to explain how things happen. This can be interpreted as a strong indication that more lies behind those happenings than the purely physical. This is of course anathema to many scientists, but that tends to be the result of a commitment to a particular outlook rather than an open-minded response. We entirely agree with the reluctance of scientists to introduce nebulous or vague ideas over which we have no control and which at the outset deny the possibility of clear insight based on our ordinary human faculties of knowledge. We have explained our approach in Chapter 1. We are suggesting here that by taking the possibility of counterspace seriously, non-physical factors may be introduced in a controlled, clear and comprehensible manner which does not commit investigators to wading into an impenetrable swamp. This may then lead to an understanding of spirit that is neither vague nor swampy.

What may differ in the view suggested here from that of some other workers in the field is the proposal that counterspace linkages may apply as well to the inorganic as to the organic world. As we shall see, there are many kinds of possible linkage, and the difference between various 'worlds' or 'kingdoms of nature' lies there. We have introduced the idea of strain and stress for very simple linkages, and in fact more sophisticated versions will prove necessary. The important feature is the idea that the two spaces may be linked and that strain and stress can result from actions or transformations of linked objects. We will describe various explorations of special cases which we are in the process of integrating into a complete picture.

We decline as yet to 'explain' a linkage or how it is made, but adopt the simple proposal that a point or plane may belong to two spaces at once without saying how. This is in line with our whole approach, which is to explore the fruitfulness of the central idea introduced in this chapter and find out how, guided by phenomena, we may learn to understand the way it applies to the world of our experience. Later we will see that linkages may be understood less abstractly.

4. Counterspace Strain

It is valuable to review the concept 'force,' as our thesis depends critically upon it. When Sir Isaac Newton laid the foundations for classical mechanics, he related force to rate-of-change of momentum, momentum being the impetus of a moving object. The faster an object is moving, or the more massive it is, the greater its effect when it hits something. Thus a feather moving at 40 mph will do little if any damage to a window while a brick travelling at that speed will break it. On the other hand a brick travelling at a tenth of one mph may not break the window whereas a feather travelling at 100,000 mph will. Now momentum entails mass, another non-geometric concept. In his theory of gravity Newton only *described* the force involved; he was unable to explain its origin beyond saying that it is engendered by mass. Einstein went further by relating it to the geometry of space itself, but his transition from geometry to force did no more than Newton had done as far as the nature of force is concerned. Generally in physics force is regarded as arising from the change of momentum of small particles such as virtual photons, so that the attractive Coulomb force between an electron and a proton arises from the exchange of negative-mass virtual photons, which impart negative momentum to the two particles and hence cause attraction. This rests upon the undefined concepts 'mass' and 'momentum.'

Now the phenomenological approach is interesting. From that viewpoint we have no cause to suppose that forces act in nature because strictly speaking they are unobservable. We only observe motion, such as that of falling apples, and the only occasion that obliges us to introduce the idea of force is when we are personally involved, for instance, in pushing a heavy object or suffering the impact of one. Thus only when our will is involved do we experience force. Then we argue by analogy that since *we* have to exert force to change the state of motion of a body, then force must act in nature when we observe objects change *their* state of motion. But that is only an analogy, for we do not observe such a force. It might be objected that we can, for example, arrange for a spring to deform when hit by a projectile, but all we will *observe* in that case will be a change in size of the spring, not a force. Again we would be arguing by analogy as *we* would have to exert a force to change the length of the spring. This is the one-sided phenomenological view, which is hard to refute.

What seems to follow reasonably from the above considerations is that *force only arises when action is impeded,* for instance, for the falling stone. Certainly we observe this ourselves, for we only experience strain and force when we try to alter the natural state of motion of something massive. We do not need to espouse the phenomenological view entirely one-sidedly, but its arguments do help us to see this. For a spiritual approach to physics, we are not concerned with introducing sentimental-ism or in-principle-vague ideas that are supposed to be 'spiritual.' Rather we note that the origin of force is quite beyond an explanation in phys-ics as it stands, and the concept has only been acquired in the realm of being and will, namely our own. Where we are tempted to ascribe force to events in nature, we should remember this and either adopt the phe-nomenological view or else postulate the existence of *will* also in nature, if we are to remain true to observation and analogy.

The thesis proposed in the previous chapter — that stress arises as a result of strain when space and counterspace are linked — is based on this idea, namely that when we cross over from geometry to phys-ics by invoking any of the concepts 'force,' 'mass' or 'momentum' then those concepts are only needed when actions are impeded or altered, by human beings or by other agencies. We will assume that we go beyond phenomenalism only to the extent that we do not suppose human beings to be the only sources of will. We depart from solipsism in so far as we suppose other human beings also have will, and further in ascribing will to animals, and further still if we assume other agen-cies are active where motion is otherwise altered. This will be entailed by implication in our notion of stress, and introduces a first approach to the spiritual aspects.

The concept of strain and stress in relation to linkages has been introduced through a very simple example for a cube. It turned out as our investigation proceeded that, for many purposes, that example is too simple, but it is given as the bottom rung of a ladder we will attempt to climb. Before we ascend, it is interesting that the cube will suffer no strain, and therefore no stress, if it is rotated about an axis passing through the counterspace infinity at *O*. Only two transformations permit strain-free action: a rotation, or a reflection. This may be related to the ubiquitous tendency we find in nature for solid objects, water and gas to end up in rotary motion, for we may presume nature will seek strain-free actions. This accords with our ordinary experience in physics: strain by its very nature will try to relieve itself.

The problem with our 'first-rung' approach arises from the way we think about counterspace. There is no justification for assuming that there is any 'extensive' relation between it and space, for instance, if a point is linked it does not follow that there is any correspondence between its distances from other unlinked points in space and the counterspace shifts between it and other unlinked points in counterspace. Indeed to suppose that is to assume that counterspace is like space in having a corresponding extensiveness. Instead it is more consistent to suppose that at each linked point counterspace appears *intensively*, in the sense that at that point a polar-Euclidean influence occurs that is only evident at that point. Likewise if a plane is linked then we might expect an orientation to be imposed on space, or be active in a tangential sense within the plane, but the angles between it and other unlinked planes in space are not relatable to turns between it and other unlinked planes in counterspace.

Next we recall the important role played by the infinite plane of space, and the infinitude of counterspace, if measurements are to be made. If a linkage is to have any significance both infinities must also be linked, or no 'leverage' is possible to permit effective influences to act between spaces. Forces require two points of application (fixed points or masses) to be effective, and we expect something similar for the proposed stress between the two spaces. But if the counterspace infinity is linked, then to relate it spatially to another linked point runs the same risk of borrowed extensiveness as for any other two linked points.

One way of resolving this, adopted in this book, is to regard every linked point as an image of the counterspace infinity, which we will abbreviate to CSI. George Adams first proposed such an idea when he came to see a counterspace infinity at the growing point of every shoot. Unless we propose millions of counterspaces to account for that insight, we need instead to see the intensiveness of one counterspace manifesting at every linked point, in the form of a CSI. The counterspace that is imaged in this way we will refer to as a *primal counterspace*. Now such an idea can be seen as fractal in nature, particularly for shift which — unlike turn — is scale invariant. The notion of intensiveness is well captured by seeing an infinity at a linked point, for the spatial point is at infinity for counterspace and the rest of the counterspace is 'inward' away from the point.

To explain what is meant here by 'fractal' we will take the so-called 'chaos game' as an example. The diagram below (Figure 8) shows a fern-like form constructed using the Collage Theorem (see Gleick 1987, p. 236 for a popular account):

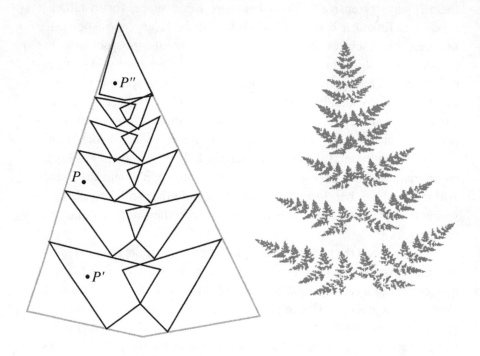

Figure 8. A fern-like construction using the Collage Theorem.

The figure on the left shows the 'tiling' used to construct the fern on the right. Each 'tile' (a shape similar to the main form) actually represents a transformation made up as follows: a translation, rotation and contraction. Thus given any starting point P, P is regarded as belonging to the main (largest) form or space, and then one of the tiles is chosen at random (the bottom left here) and the corresponding point P' in that tile is related to it as P is to the main form. P' is then regarded as belonging to the main form, and another tile is chosen randomly (the top one here) to give P'' in a similar manner, and so the process runs. The initial 50 points are suppressed to enable the process to 'settle,' and the result is the fern form on the right. The random choice of a tile on each pass has nothing to do with the fractal, but is purely a Monte Carlo approach to obtain a complete picture in a reasonable time. The point of this is to see that the resulting form arises from a number of simultaneous spatial transformations, the surviving points on the fern being those which satisfy them. The Collage Theorem states that, if certain conditions are satisfied, this process yields a genuine fractal form.

We regard a number of CSIs linking the same counterspace to space as being analogous to the above tiles, the *primal* counterspace being associated with the main form. However, we do not regard the primal infinitude as itself being linked other than via the CSIs. This is only intended as an analogy to illustrate what is meant here by a fractal coupling of space and counterspace.

The kind of strain that arises for this sort of linkage is different from that for the simple case of the cube. We will refer to it as *affine strain*, as there are different ways in which space and counterspace may be linked, depending upon how 'strict' the linkage is. There are different underlying geometries for each type of linkage, one such geometry being so-called affine geometry. A major characteristic of that geometry is that direction in space is well defined but distances and angles are not, so we say direction is an affine concept. We experience affine geometry every day through vision in perspective. When viewing an avenue of trees the size of the trees appears to diminish with increasing distance even if they are all in fact the same height. Also their distance apart similarly appears to decrease. Thus our *direct visual experience* is not 'metric,' that is things of the same size do not appear to be so. But direction *is* well defined because when we move around we see that directional relationships are unchanged. Now a direction is related to infinity, because two lines point in the same direction if they are parallel, and parallel lines meet in a point at infinity. Ordinarily we say that two parallel lines do not meet at all, but mathematicians have found in the last three centuries that it is valid to add a special or 'ideal' point where they are said to meet, which turns out to be infinitely far away. Similarly two parallel planes meet in an ideal line at infinity, noting that two planes intersect in a line. Since directional relationships are unchanged, it follows that the sum total of ideal points and lines remains unchanged in affine geometry. Now a set of lines all meeting each other in diverse points must lie in a plane, so we say there is a 'plane at infinity' which affine geometry leaves unchanged. When we take the polar opposite of this in counterspace then points are polar to planes, so we have 'parallel points' lying on a 'line at infinity,' which is a line containing the infinite point of counterspace. We see here an example of how different counterspace is from ordinary space, posing quite a challenge to our space-based intuition! This will be of great importance when later on we come to consider light, where 'polar affine geometry' is involved. This is explained further in Chapters 6 and 7. The concept of affine strain is illustrated in Figure 9:

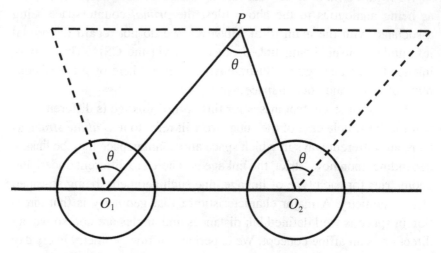

Figure 9. Affine Strain.

Two CSIs are shown as the centres O_1, O_2 of two circles. The real meaning of the fractal quality of the linkage is that the primal counterspace sees P in different directions as viewed from O_1 and O_2. There is a directional strain indicated by the angle θ. If the linkage is metric then this angle represents a shift strain, but if the linkage is affine then shift is undefined and it indicates a directional incompatibility which we are calling affine strain, as direction is an affine concept. If P moves away from the line O_1O_2 then the angle θ decreases and the strain decreases, and clearly if P goes to infinity in the Euclidean sense then the strain becomes zero. As we are proposing that point linkages are CSIs, P will also be a CSI, and we see that the effect of affine strain is to cause expansion. This assumes that O_1 and O_2 are fixed and P is free to move. If all three CSIs are free to move then the situation is somewhat different and we shall see how to analyse that shortly.

There will only be movement if the strain gives rise to affine stress, and we need to see how the two are related. The problem with angles is that they are not vectors, and it is preferable to work with vectors (see Chapter 5). An example of a vector is velocity. A car moving at 80 mph is doing so in a definite direction. Its *speed* is simply 80 mph, which is all that will interest the police, but the combination of speed and direction of travel is *velocity*. It requires energy to change the speed, and it also requires energy to change the direction of motion. Consider a steam train travelling north at 60 mph with an east wind of 10 mph, then to solve the old problem of finding in which direction the smoke goes requires the use of vectors, not just speeds. If instead we say the train is travelling at 96.6 km/hr and the wind is blowing at 16.1 km/hr and we say the whole thing in German, the observed physical result is unchanged. *The vectors do not depend upon how we describe them.* This important property of vectors is of greatest importance in physics, which is why we wish to use them.

Returning to shift strain, for our purposes the simplest approach to affine stress is to consider the rate of change of the strain, for rate of change of angle is a vector. We thus propose that:

Affine stress is proportional to rate of change of affine strain.

In the metric case where we have shift stress, rate of change of shift is also a vector. The idea is analysed in detail elsewhere (Thomas 1999, p. 49) where the 'chord law' of affine strain is derived (Figure 10, and see also Chapter 6):

If three CSIs interact through shift or affine strain, then referring to their common circle, the magnitude of the strain gradient at a CSI is proportional to the opposite chord over the product of the adjacent chords, directed towards the centre of the circle.

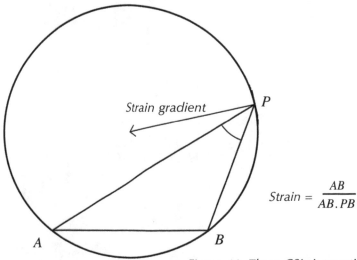

$$Strain = \frac{AB}{AB.PB}$$

Figure 10. Three CSIs interacting.

The resulting stress requires a transition from geometry to physics as we have seen, and the actual value of the stress cannot be derived from the geometry alone as it depends on the physics, for instance, whether we are dealing with a gas or liquid or whatever. Thus the magnitude of the stress must be *scaled* to the strain by a non-geometric parameter. In the case of a strained metal bar, for example, Young's Modulus which relates stress to strain is 210 GPa for steel (1 giga-pascal is 145 pounds per square inch, or about 10 times the pressure of the Earth's atmosphere) and 117 for copper, as the same strain in each case results in a different stress. These constants have to be found experimentally. In an analogous way we will need to allow for *scaling* in the way space and counterspace are related. Scaling determines whether a large change in space corresponds to a small one in counterspace, or *vice versa*.

In this chapter we have progressed from a simple case of strain to a more sophisticated situation based on the fractal-type linkage proposed. As we proceed we will come across other kinds of linkage than the point-wise one introduced here. In the next chapter we will consider gravity which was the first problem treated by this approach.

5. Gravity

Gravity is a rather mysterious force. We are born into a world where we experience it from infancy onwards so that the notions 'up' and 'down' and the fact that unsupported objects fall down is so natural and obvious that it took the genius of Sir Isaac Newton to recognize that there must be a scientific reason for this state of affairs. He proposed that massive objects exert a force on each other which depends upon their mass and separation. It is said that he first thought of this when seeing an apple fall, conjecturing that the Earth exerts a force of attraction on the apple. He then applied this idea to the Moon, supposing that the Earth and Moon attract each other for the same reason, resulting in the nearly circular orbit of the Moon round the Earth. The reason it orbits rather than behaving like the apple is that it is moving so fast that although it does fall its velocity is sufficient to compensate for that fall and keep it in orbit, so that it falls 'round' the Earth instead of on to it. Hence the term 'free fall' in space science.

However, Newton felt uncomfortable with the idea of 'action at a distance' which his idea seems to entail. We know that we can exert force by pushing or pulling an object, but this requires direct contact with the object concerned. There is no obvious direct contact between the Earth and the Moon, so action at a distance seems unavoidable. But another approach is to suppose that a massive object distorts the space around it so that other freely moving nearby objects follow curved paths instead of straight lines. If their velocity is not too great and the primary object is massive enough then those curved paths may be closed in the form of circles or ellipses, and thus the objects end up in orbit. This idea can be applied to the Moon and the Earth as well as to the planets moving round the Sun, and is due to Einstein (it is a consequence of his General Theory of Relativity). Of course the concept of curved space is hard to grasp as we usually think of it as featureless and structureless, but earlier chapters should have made clear that such an idea is too simplistic. Einstein tends to explain gravity geometrically, but the distinction between geometry and physics must still hold good (compare previous chapters) as he did not explain 'mass' geometrically, and of course it plays an essential role in his theory. Essentially, then, he proposes a subtle relationship between mass and space itself, and deduces gravity from that. This relationship remains unexplained.

We will approach gravity in a new way based on the ideas of the previous chapter. For this we will concern ourselves with point linkages between space and counterspace and find the kind of stress that arises from the shift strain we introduced previously.

We propose that we have a sphere with a counterspace infinity at its centre O, which is interacting with another CSI at S (see Figure 11).

P is a point of the sphere (which we assume is solid). The factor we rely upon is that the point P is 'seen' by a primal counterspace in a definite direction from O, but in a different direction from S. Now if S and O relate to the same underlying primal counterspace, as we proposed when discussing shift strain — so that they are really images of its infinitude at the point of linkage — then it will have two 'eyes' as it were at S and O, but P will be inconsistently linked from these two views. We proposed in the previous chapter that this gives rise to shift strain. We are concerned here with what look like angles, but for counterspace they will be shifts as these angles relate to the CSIs and so may be seen as measures of shift. The shift strain is represented by the shift equal to the angle SPO.

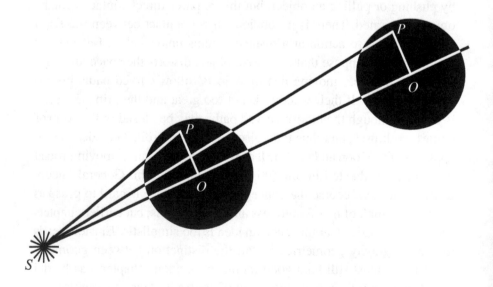

Figure 11. Shift strain in spheres in counterspace.

If the sphere moves towards *S* then the angle *SPO* gets smaller, so we see that the strain will be relieved as a result, and thus we suppose that an attraction is to be expected to reduce the strain. This, when developed in more detail, is the basic idea of gravity we are proposing. So far we have only considered strain, but if a force is to arise we must move from geometry to physics and see what kind of stress is related to this strain.

It is necessary to discuss a most important fundamental idea before we can proceed further, which has to do with the fact that we cannot expect nature to be in the least concerned about the way we describe her. What this means is that if we set up a system of measurement in which we choose a zero point and measure angles and distances — or shifts and turns — in relation to it, we can deduce formulae describing what happens based on such measurements. What we would not expect, and do not want, is that the choice of a different basis for our measurements would lead to different results, for the way things behave surely cannot depend on that! It is known that some quantities are suitable and fulfil this condition while others do not, and mathematicians have very ingeniously developed a branch of mathematics which tells them which kinds of quantity fulfil our requirements. One outcome is that angles are not suitable, and neither are shifts. An example of something which is suitable is a velocity, so that if a car is travelling due north at 50 mph then it does not matter whereabouts we set up our radars or other instruments by the road to measure that velocity, they will give the same result. This is clear as velocity is a quite definite rate of movement with respect to the ground which is obviously unaffected by how we measure it if we do the job properly. A quantity like velocity which has both a magnitude and a direction (we specified the direction of travel above) is called a *vector*, and there are many examples of vectors in physics.

It is equally clear that the force of gravity should similarly be unaffected by how we measure it, as it also is an independent quantity, and indeed the force that arises if we prevent the fall of an object is a vector, for it has a magnitude measured in kilograms and a direction, that is, downwards. Thus in order to relate the shift strain described above to stress, we require a vector. Another quantity that fulfils the above requirements is the rate at which an angle is changing. Thus if we say that a wheel of a car is rotating at 800 rpm then we are referring to a rate of change of angle, for each revolution is a change of 360 degrees and we say this happens 800 times a minute. The vector is conventionally represented by a line along the axle with a length proportional to the rpm and a direction depending upon whether the rotation is clockwise or anticlockwise. Similarly we find that the rate of change of shift is a vector, and is more suitable for our purposes than shift itself. Thus if the sphere moves towards S then the angle SPO changes, and the rate of change of this angle is a measure of the rate of change of shift. Now this quantity exists even if the sphere is prevented from moving, just as the gradient of a hill exists even if we are not moving up or down the hill. Indeed such a rate of change is technically referred to as a *gradient*, based on this analogy.

In our investigations, the idea was thus explored that the stress arising from the shift strain is related to this gradient or rate of change of shift. When the idea is analysed mathematically, it turns out that we obtain the correct law for gravity (Thomas 1999, Chap. 6). Newton's law is expressed by the formula

$$F = G M m / r^2$$

which says that if two objects such as the Earth and Moon attract one another then the force F is directly related to their masses M and m multiplied together, and is inversely related to the square of the distance r between their centres (the so-called inverse square law). Thus the greater the masses the greater the force, and the force also increases as they get closer together. The number G is the gravitational constant which relates the way we measure mass and distance to the way we measure force. The analysis of shift gradient (Thomas 1999, Chap. 6) shows that the inverse square law is obeyed by the stress, and also it is proportional to the volume of each body. It is also shown there that if the stress depends in addition upon the density of the body then we obtain mass as in Newton's law, instead of volume. This relates to an important topic we will come to later concerning the way space and counterspace are scaled with respect to each other.

A challenging aspect of the above results is that it is not necessary for the infinitude S to belong to a massive body; it need only belong to a sufficiently 'strong' counterspace. By 'strong' we mean that the scaling is such that density arises for the sphere in Figure 11, and hence mass. Thus in principle the Sun, for example, need not be a massive body to hold the planets in orbit. This throws light on Steiner's discovery that the Sun is not a 'ball of burning gas' but is a counterspace. It can be so and yet interact gravitationally with the planets.

In conclusion, gravity may be described as the rate-of-change of shift strain between space and counterspace.

6. Elements

The analysis of gravity in the previous chapter and its success in deriving Newton's law was the first indication that our thesis about strain and stress between space and counterspace had any merit. Further developments were that gases and liquids can also be described in terms similar to that used for gravity, and then the work extended to light in some detail, solving the principal problem posed by Steiner's research in the process (see Chapter 2 concerning the velocity of light), and then on to chemistry and life. In other words the whole enterprise opened up much more than originally expected, starting to embrace the whole of science, albeit in a very introductory manner. But we cannot run before we can walk in such affairs, and there is an enormous amount of detailed work by the scientific community that needs eventually to be accommodated. All of this depends upon closer attention to different kinds of space apart from those we have considered so far.

We have introduced the idea of counterspace as a metric space in which measurements of an unusual kind are possible, and we recall that all our introductory examples, quite deliberately, concerned solids. So far, so has our analysis of gravity. Now in describing the properties of a Euclidean transformation we mentioned that a pint of beer unfortunately does not increase to two pints if we move fast enough. However, it is immediately obvious that the liquid can change shape, even get spilled if the train we are sitting in jerks. Although it is in Euclidean space it obviously has the freedom to flout Euclid in many ways. Recalling that this is said in the technical context that a strictly Euclidean transformation conserves all angles, lengths, volumes and areas, we must now take on board the fact that liquids do not behave like that except as far as volume is concerned. Of course we can increase the volume of our pint of beer by heating it up, but that is generally considered undesirable by the connoisseurs and we are assuming that the temperature is fixed. A very useful application of the law that liquids have a fixed volume lies in hydraulics. When you put your foot on the brake pedal of your car the brake fluid transmits the force you exert to the brake pads because it is not easily compressed to a smaller volume. Should you get air in the brake fluid then the brakes fail because the air is quite happy to change its volume instead of dutifully passing on the force.

We see then that liquids can transform in all manner of interesting ways provided they maintain constant volume: angles can change,
lengths can change, and so can area. If you start off with water in a
spherical bowl and then pour it into a long thin cylindrical mug the
surface area increases. Thus we need another kind of transformation to
handle liquids. Besides the volume requirement, liquids also share with
Euclidean transformations the fact that infinity stays put. Now it turns
out that there is a species of geometry well suited to this kind of behaviour which is known as *special affine geometry*. We will not go into any
detail in this book, but we will see that it is part of a definite sequence
of possible geometries useful for our purposes. What is important for
us is that it has the two major properties already mentioned. The word
'affine' means a relative by marriage, and came into use in geometry
when two different aspects of Euclid's work were disentangled, namely
its absolute aspects which we have already seen, and others that are
more relative, which were indeed related and 'married' in Euclid! This
geometry is actually very familiar to us as we encounter it daily through
the fact that we see the world in perspective. Some may object that it is
projective geometry that is involved there, being concerned as it is with
the laws governing projection. While that is also true, the difference lies
in the fact that in everyday life infinity mercifully stays put, and that
is precisely the difference between projective and affine geometry. So
affine geometry is a species of projective geometry that is restricted by
the fact that infinity stays at Euclidean infinity, whereas there is more
freedom in projective geometry. Hence in proper perspective drawings
we have to pay attention to vanishing lines which are fixed, whereas in
pure projective geometry they are not really significant. We used the
term 'special affine geometry' because it constrains volume to be invariant whereas more general affine geometry allows volume to change, but
still respects infinity.

Another important property of affine geometry is that parallel lines
remain parallel, as do parallel planes. In other words an affine transformation will transform two parallel lines into two new lines which are
again parallel, and similarly for parallel planes. This is clear if we think
of the ideal point in which two parallel lines meet. It is in the ideal plane
at infinity, and since that plane is invariant as a whole (as affine geometry leaves infinity fixed), the transformation must move the ideal point
into another ideal point still in that plane. Hence the lines are moved by

the transformation into two new lines which meet in the new ideal point, and thus they are also parallel. Likewise parallel planes meet in a line at infinity, and that line must move to a new line at infinity, and the corresponding two new planes must meet in it and hence are parallel.

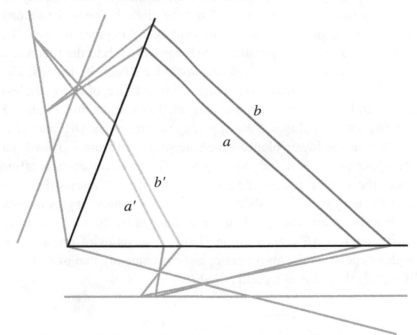

Figure 12. Special affine transformations maintain parallel properties.

Figure 12 shows two parallel lines *(a* and b) transformed into two new ones *(a'* and *b').* But, two lines that are not parallel but at some other angle are moved into two new lines which may be separated by a different angle: angles are not preserved by affine transformations, and neither are lengths. We have the extraordinary and important fact that we *cannot compare lengths of lines that are not parallel.* This is because the ratio of their lengths is generally different after the transformation, so no standard of measurement can be set up. However, and this is crucial for our later study of light and also of gases, line segments that are parallel have a definite ratio, and after the transformation the new segments will be parallel as we have seen, *and in the same ratio as before.* Thus if one segment is twice as long as the other then its corresponding segment after the transformation will be twice as long as the other transformed line segment.

Something similar is true for angles: suppose that we have a plane containing two lines at 30 degrees and a parallel plane containing two other lines at 60 degrees. The ratio of those angles is 2. After the transformation the new planes will still be parallel, but the two new lines of the one plane may be at some other angle than 30 degrees, say 40 degrees for the sake of argument. Then the angle between the second two (new) lines in the second plane will be 80 degrees because the *ratio of angles* in such a parallel case is conserved. Had the first angle been 20 degrees then the second one would have been 40 degrees. The same holds good for areas. So two areas (for instance, of two triangles) that are parallel have their ratio conserved. To illustrate this, Figure 13 shows two parallel planes. Two superimposed triangles ABC and ADE are shown in the lower plane, which are projected from a point P on to the upper plane as $A_1B_1C_1$ and $A_1D_1E_1$. The transformation is affine because the planes are parallel, so the ratio $AB{:}AD$ equals the ratio $A_1B_1{:}A_1D_1$, and $AC{:}AE$ equals $A_1C_1{:}A_1E_1$. Also the three angles in each triangle retain their ratios and so are unchanged by the projection, for instance, angle CAB equals angle $C_1A_1B_1$. The ratios of the areas of triangles depend on the above ratios and consequently that of ABC and ADE equals that of $A_1B_1C_1$ and $A_1D_1E_1$.

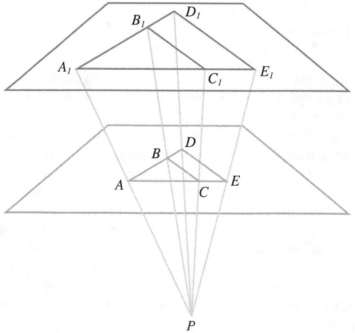

Figure 13. Transformation in affine geometry where ratios of angles and distances are maintained.

The practical consequence of this is very important, for it means that in affine geometry we can only make comparisons of lengths or areas if they are parallel, and of angles if they are in the same or in parallel planes.

As you may have feared, there is a polar truth to this for counterspace which is of great importance. We can only compare polar areas if they are 'parallel' in counterspace. In Chapter 2 we saw that the form polar to a circle is a cone. In affine space we can compare the areas of circles if their planes are parallel, and in polar affine space we can compare the polar areas of cones if their vertices are 'parallel,' that is, lie on a line through the CSI, or of course if they coincide. This will prove very important for the study of light.

After this necessary digression, it is perhaps now easy to take another step and relate affine geometry to gases in addition to relating special affine geometry to liquids. As volume is not conserved in pure affine space it permits expansion which is one of the notable qualities of gases. Thus we now have the scheme:

metric geometry   solids
special affine geometry   liquids
affine geometry   gases

The final step would be to relinquish the constraint on infinity to give us projective geometry. It turns out to be too simple to try to relate that to heat because measurement is still possible in relation to heat but projective geometry does not permit that. However we do see now a very interesting correlation between different kinds of geometry and the traditional elements: earth (solid), water (liquid) and air (gas). Fire (heat) will occupy us later.

Since affine geometry does not conserve angles, we realize at once that if we are to study gases and liquids in a way similar to our procedure for gravity we have to make do without shift and shift strain. But that does not mean we can never use angles, for the following reason. Recall that gases and liquids exist in Euclidean space, but within its confines they transform more freely. We can mirror this fact by supposing that they also are linked to both space and counterspace — those spaces remaining as we have described them — but the linkage is not metric but affine for gases and special affine for liquids. This enables us to use angles in space to measure strain even if those angles are changing. Referring back to Figure 10, recall that the primal counterspace 'saw' the point *P* stereoscopically from the CSIs *A* and *B* which gave rise to direction strain. Because we were still working in a metric context we described that as shift strain and used it later for gravity. In an affine context we do not need to use the concept shift but can stay with the fundamental fact that we have a directional incompatibility causing strain, and in our present context we call it *affine strain* (compare Chapter 4). Because the affine linkage is embedded in a Euclidean context we can still use the angle *APB* as a measure of this strain. The difference from the gravity case is that the relative positions of *A, B* and *P* are now free to move within affine constraints, and we then seek the maximum rate at which the affine strain can change to give us a suitable vector. The analysis of this gives a remarkably simple law. Imagine that we leave *A* and *B* fixed and move *P* around quite freely in the plane of the diagram. If we draw a circle through *A, B* and *P* and imagine *P* moving along a tangent to the circle we find that the angle *APB* is momentarily unchanging, in which case the strain is also unchanging (see Figure 14).

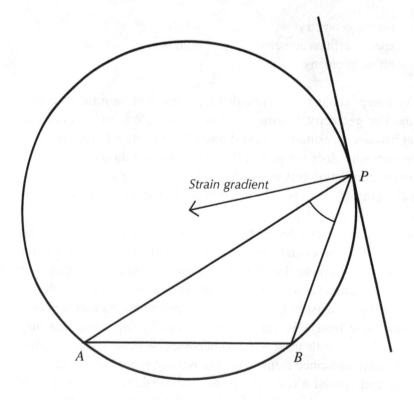

Figure 14. Affine strain.

On the other hand if we move P along a line through the centre of the circle to P'' the angle clearly changes quite rapidly, and it turns out that in this particular direction it is changing more rapidly than in any other. This maximum rate of change is called the *gradient*, and it is a vector as it has a definite magnitude and direction. It is thus suitable for us to use as a measure of the stress arising from the affine strain, and we call it *affine stress*. Thus we follow a similar path to that taken for gravity in pursuit of a consistent approach to the way strain and stress are related for the elements. When we calculate the magnitude of the gradient mathematically we find it is proportional to the quantity

$$\frac{AB}{AP \times BP}$$

which is a remarkably simple and convenient law which we refer to as the *chord law* of affine stress. This is because the lengths AB, and so on, are chords of the circle through the three points. What this shows is that if AB stays fixed and P moves away then AP and BP increase and thus the stress decreases, and the strain also decreases since the angle APB gets smaller. We thus have nicely mirrored in this the expansive quality of gas, and we have a 'levity'-like quality which is the opposite of what we found for gravity.

We apply this law to study the behaviour of gases. We imagine we have a spherical container with gas in it. Now we come to an important discussion of what we think a gas is, as our application of the law may otherwise lead to misconceptions. We will regard the gas in the container as composed of a large number of CSIs, but that does not mean we are taking an atomistic view in the usual sense. The conventional picture is of many molecules flying about inside the container suffering collisions with each other and in particular with the walls. A molecule is supposed to be a very small object with mass and in a gas it is moving and thus has momentum. The hotter the gas the greater the average momentum (and energy). When it hits the wall it usually bounces off and thus changes its direction of motion, so its momentum changes which entails a force which must be supplied by the wall. This is happening continuously all over the inside of the container as molecules hit it and bounce off, so the whole container experiences forces all over its inner surface, which sum to give the pressure of the gas. If we increase the volume of the container then its inner surface area increases, so the collisions are distributed over a larger area and thus the pressure, being an average, is reduced. Thus for a fixed temperature, and therefore average momentum change, the pressure and volume vary in opposite senses which is known as Boyle's Law, expressed by the equation PV = constant, where P is pressure and V is volume. The kinetic theory of heat supposes that an increase in temperature throughout the gas results in the molecules moving faster on average, and thus the average change of momentum resulting from collisions increases, and hence the pressure increases. Thus pressure is proportional to temperature. This gives the *ideal gas law* obtained by combining this result with Boyle's Law to give the equation $PV = kT$ where T is the temperature and k is a constant. It is called the ideal gas law because it applies in ideal circumstances to a gas with ideal properties. In fact it works well over a wide range of conditions but is inaccurate in certain special circumstances. This law can be experienced when a bicycle pump gets hot because we keep reducing the volume and increasing the pressure of the air in it as we force the air into the tyre. Clearly the pressure increases more than the volume decreases since the temperature rises.

This is the conventional view of the behaviour of ideal gases. The law is well tested and we should be able to derive it based on our counterspace premises, and indeed we can. However we will not be assuming that the gas consists of molecules moving very fast and hitting the wall, but instead we regard it as consisting of many CSIs experiencing affine stress which we have seen is expansive and thus the pressure arises for quite a different reason. The CSIs are affinely linked which means that their positions in the gas are not accurately defined other than at the walls where the metric quality of the container is imposed on CSIs in contact with it. Thus where the kinetic theory sees little masses moving fast we see CSIs with fuzzy locations. We thus understand why, from our different perspective, the kinetic theory works: it images the real situation sufficiently well to be a good model, even if not the truth. This was a point Steiner was emphatic about, and we have been able to approach it not 'head on' as an exercise in apologetics, but as a result of consistently following up our new paradigm.

When we apply the chord law we do so to CSIs in planes as we are concerned with an affine situation which only permits comparisons of quantities in the same direction (see the discussion of affine ratios above), and since the affine strain is suffered by counterspace, 'direction' means 'in a plane.' We then sum the contributions to the total affine stress of all possible triangles of CSIs in the gas (such as in Figure 14) which have two CSIs on the wall. This ensures we always stay in planes and cover the whole volume. We get a definite total affine stress distributed over the inside surface of the container giving rise to pressure. If we now increase the radius of the container then all the triangles increase in size, and so according to the chord law the affine stress caused by each triangle decreases proportionally. This is because on average the areas of the triangles increase linearly so the stress decreases linearly with increasing radius. Hence we obtain Boyle's Law quite simply when we combine this with the fact that the pressure is distributed over the inner surface. When that increases it causes the average affine stress per unit area to decrease and hence the pressure falls. The volume dependency arises because we are combining the effect of an increasing area with a falling average shift stress proportionally to the radius, and area multiplied by radius is proportional to volume. The chord law also ensures the affine stress makes the gas expand to fill the container so that the average number of CSIs near the wall is constant, and hence also the number of triangles involved (see Figure 15).

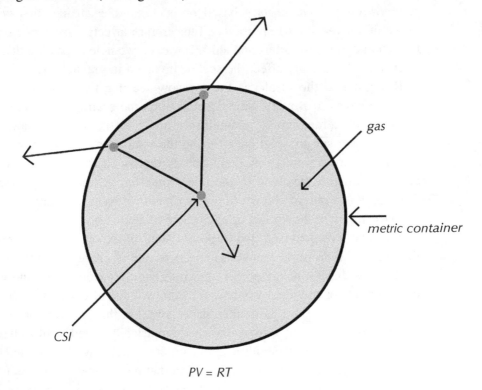

$$PV = RT$$

Figure 15. The ideal gas law.

The temperature aspect now leads us to an important point which we must discuss before we can finally arrive at the ideal gas law itself. Up to this point we have quite uncritically assumed a simple relation between space and counterspace such that turn, for example, is related to the reciprocal of radius. But do we say the turn is related to $1/r$, or $10/r$, or what? There is no particular reason why it must be $1/r$ since the two spaces are disjoint rather than 'back to back.' This concerns the *scaling* between the two spaces. It could well be that a small change in spatial radius is accompanied by a very large change in shift, or *vice versa*. Notice that it is more correct for a non-extensive linkage to think in terms of rates of change being compared rather than shifts and distances directly. Now consider a solid block of iron, which we assume is linked metrically to counterspace. If we heat it up then its size changes, which is not a metric transformation! Thus the application of heat changes its relation to counterspace and by implication its scaling to counterspace. We now begin to see why we could not simply relate heat to projective geometry. It is a complicated subject mathematically and we will not pursue it far in this book, but we specifically postulate that *scaling is proportional to temperature* based on our example, an idea that will prove of far-reaching importance. Temperature affects most aspects of the linkage, not just radius and shift. However, we note in passing that it does not substantially affect the scaling involved in gravitation.

Returning to the ideal gas law we now see that the way we relate the shift stress to Boyle's Law depends upon the scaling, and that will change with temperature, so finally we arrive at the ideal gas law. In particular we see the scaling affecting the relation between strain and stress, as temperature is another quantity that is clearly not geometric and is intimately involved in the transition from geometry to physics. The stress is a rate of change (a gradient) and we have seen that scaling is concerned with how rates of change are related.

We have now achieved two important steps based on the strain/stress thesis and the proposal that stress relates to rate of change (or gradient) of shift: we have a new theory of gravity giving Newton's Law, and we can account for the ideal gas law. We acknowledge of course that there is more to gravity and gases than these laws, but this is a very promising start which gets to grips with actual empirically tested results rather than remaining with generalities. Much more detailed work is needed to take the subject further in these areas, but that must come as opportunity and insight (and the necessary work) permit. Such work is actively in hand in many other areas than just these two. Our aim is to achieve first of all a global picture of science based on our fundamental strain/stress thesis, providing entries into different areas as we have done for gravity and gases.

Returning now to liquids, we follow a similar route but with special affine linkages instead of general affine ones. We need to capture the constant volume property, and for this we use tetrahedra instead of triangles. It will be recalled that a tetrahedron has four vertices (that is, points) and four faces and each face is a triangle. We can apply the chord law to the three CSIs in each face and see what effect that has while the volume is held constant. Each CSI belongs to three triangles and thus the total affine stress it experiences is the sum of the stresses on it due to the three triangles, allowance being taken of the directions while the sum is taken (a vector sum). Then the component of the total stress which does not tend to change the volume is what will move the CSI as the tetrahedron changes. The mathematical analysis of this is quite intricate and we will content ourselves here with some of the results:

1. A single tetrahedron is in balance when it is regular, that is when its edges are all of equal length.
2. A tetrahedron in isolation ends up as a regular tetrahedron, unless it is flat (of zero volume).
3. The stress is inversely proportional to the linear size of the tetrahedron, by which we mean that in the case of a regular tetrahedron it is inversely proportional to the length of its edges.
4. The stress is never zero, so the liquid remains 'sensitive' in a dynamic balance, unlike a solid.
5. A 'flat tetrahedron' behaves chaotically.
6. A tetrahedron with a small triangular base and its fourth vertex relatively far away acts to pull the base towards that vertex (not the other way round as might have been expected).
7. For a tetrahedron with two vertices close together and near the centre of the line joining the other two vertices (which are much further apart), the first two vertices rotate quite rapidly about the longer line and the other two very slowly move inwards on a spiral.

These results have some interesting qualitative implications quite characteristic of liquids. Result 3 shows that there are long range forces acting in the liquid, not just short-range ones, their strength falling off proportionally to the range. Result 2 shows that long thin tetrahedra try to become regular and so the liquid will try to reduce the ratio of its surface area to its volume, tending to form a sphere. A blob of mercury on a surface best illustrates this effect. However if the liquid forms a thin film on a surface then the greatest forces act on the smallest tetrahedra, so short-range effects dominate, and we see why a liquid like water might tend to form temporary short-range tetrahedral structures. The smaller they are the longer lived they are likely to be, leading to microstructure. This depends, however, on the scaling between space and counterspace for the liquid in question, which will affect how strong the affine stress is, and if it is weak, microstructure need not arise. Then it will tend to be more volatile.

Result 6 is most instructive. Imagine a drop of liquid. Near the centre there is equilibrium, but in the surface there are many small triangles forming many long thin tetrahedra with CSIs inside the drop, and all of these will strive to pull the triangular bases in the surface inwards, so we see why the drop tends to become spherical, and why there is surface tension.

Result 5 is also interesting in two ways. First of all it accounts for Brownian motion based on CSIs rather than collisions of molecules. This motion was first observed by Robert Brown in 1827 who saw small pollen grains suspended in a fluid suddenly jump to new positions as if hit sharply by something. The effect appeared to be quite random and such fluctuation of small particles has been observed in many other circumstances. Very small 'flat tetrahedra' may be expected to produce this as they bring in a chaotic tendency with sufficient force if they are small enough.

The second aspect of Result 5 concerns the surface of a liquid. In three-dimensional special affine geometry volume is conserved but not area. Thus the liquid surface should be 'gas like' as it behaves affinely rather than special-affinely: hence we have the gas-like behaviour called evaporation. The chaotic motion of 'flat tetraheda' in or near the surface will tend to be more pronounced as they are not surrounded by other CSIs on all sides, and may contribute to the process of evaporation.

Result 7 indicates a source of vorticity in a liquid. Liquids tend to develop vortical motion quite readily and there will be many such long tetrahedra in a liquid mass.

If we have two liquids in the same container then hybrid tetrahedra will exist made of CSIs from both liquids. The resulting affine stress may be stronger or weaker than for a homogeneous tetrahedron. If weaker we expect the two liquids to be immiscible (like oil and water), but if stronger then they should mix readily (like detergent and water).

This also raises the question about hybrid tetrahedra formed between the liquid and the walls of the container. Such details throw light on capillary action and for gases may, when properly analysed, lead to some of the corrections to the ideal gas law that have been found necessary and described by van der Waals.

Thus we see that application of our postulates in a first relatively simple analysis of liquids yields several qualitative results characteristic of liquids, encouraging further work and reinforcing the consistent approach taken to shift- and affine-stress.

In this chapter we have concentrated on pointwise linkages based on CSIs where the stress is suffered in counterspace rather than space. This has thrown light on the aggregates of matter traditionally known as the four elements, but although a first approach to heat has been made via temperature in relation to scaling, there is more involved to which we will return later. Our results have critically depended upon the notion of the fractal linkage of space and counterspace. We have not attempted to enter into the details of fractal mathematics, but that can be followed up elsewhere (Thomas 1999, Chap. 5). However the 'stereoscopic' relation of a primal counterspace to its CSIs is what is needed to grasp the import of this for our work.

Implied in all this is the notion that in matter we have pointwise linkages via CSIs, which explains why the point-based particle-approach adopted in physics and chemistry is so successful. It also shows why the atomic hypothesis is so successful, but we do not regard CSIs as material atoms as their very definition shows.

We have concerned ourselves with affine linkages, but it is clear that counterspace can also have affine transformations relating to turn and shift, which we will refer to as *polar affine transformations*, and we will turn to this subject next when we study the ethers in polar contrast to the elements.

7. The Ethers and Light

So far we have looked at the possibility of point-based linkages, and it might be imagined that now we should examine plane-based ones. However a simple approach based on linked planes as images of the spatial plane at infinity has not proved fruitful. One reason lies in a property of shift known as *scale invariance*. If we recall the chord law it applies regardless of the size of the triangle, that is, the scale does not matter (we are not using 'scale' in the sense of the scaling between space and counterspace, but as it is related to size). This is an important property of fractals which are precisely concerned with structures which look the same on any scale, for instance, a fern leaf appears to be made of smaller fern leaves. Shift- and affine-strain are essentially fractal in nature and rely on this. But turn is not scale invariant and a fractal approach has not been successful. We expect strain based on the planar aspect to relate to something polar opposite to the four elements, and this we will indeed find. Steiner discovered in his research that there are four subtle non-sense-perceptible aspects of the world which he called 'ethers.' He related them to *warmth, light, chemical action* and *life itself.* However the chemical ether is also referred to by him as the *number ether* and the *tone ether*. The work to be described in a later chapter throws light on why this particular ether should have these diverse characteristics, and indeed it was an outcome of *not* merely assuming what Steiner said, but following where the ideas on counterspace lead, that clarified this. In our investigations, however, a 'head-on charge' based on merely taking the polar opposite of what worked for the elements did not work. A more subtle approach will become apparent as we proceed.

On one occasion Steiner reported that ether is essentially two-dimensional, so we must be careful not to identify 'ether' with 'counterspace' which is three-dimensional. That is not to say that there are no two-dimensional counterspaces, but so far we have not been concerned with them. Before going further into the nature of the ethers we will relate how our enquiries led up to them.

Light ether

If we start with light we have a fascinating area of study to contend with. It has posed the greatest riddles to science as simple pictures of its nature are elusive. Newton thought light consisted of streams of tiny particles ('corpuscles') travelling in straight lines, and went a long way in explaining phenomena on that basis. But some experiments suggested that it consists of waves rather than particles. For example it spreads after passing a sharp edge. If it consisted of tiny particles then an obstruction with a straight edge should result in a sharply defined shadow on a screen beyond, yet that is not what happens and the edge of the shadow is more fuzzy than would be expected. This can be compared with the way waves in the sea spread out after passing a harbour wall. A very thin film of oil can produce beautiful colours, a phenomenon more easily explained in terms of waves rather than particles. If a lens is laid on top of a flat mirror and specially prepared light is shone on it then a series of dark and bright rings appear, an effect referred to as 'Newton's rings' despite the fact that it strongly supports a wave theory rather than his corpuscular theory. Then Einstein obtained his Nobel Prize not for his Theory of Relativity but for explaining the photo-electric effect. If light falls on a metal plate in a vacuum that is connected to an electric circuit, then below a threshold frequency nothing happens regardless of the strength of the illumination, but when the frequency is increased current suddenly begins to flow. This happens at a definite threshold frequency. Einstein explained this by supposing light to consist of photons each with an energy depending on the frequency (that is, roughly the colour). If the frequency is too low no single photon has enough energy to release an electron and thus generate current, but above a threshold all photons have sufficient energy so even a very low level of illumination then causes a current to flow. This seemed to support Newton's corpuscular view! Much experimental evidence since then shows that light seems to behave in both ways, particulate and wavelike, and the experimental set-up will tend to determine which 'face' it shows. The photon is not thought of as a simple 'corpuscle' but instead as a 'wave-packet.' This dualism persists to this day and now appears to be unavoidable in the current paradigm.

The first observation we can make is that because light behaves in certain ways when it interacts with matter it is supposed that the explanations based either on particles or on waves must relate to the light itself.

In other words, if light exhibits wavelike properties then it must itself be a wave. But that conclusion is not logically essential. If I lay a cloth on a grid it will then appear grid-like, but that does not mean it is itself inherently grid-like. It has the capacity to appear that way given the right circumstances. It is no more logical to insist that light is a wave because it exhibits wavelike behaviour in some circumstances. Or, on the other hand, that it is particle-like.

The second observation is that our scientific knowledge of light only arises from a study of its interaction with matter. Thus the observed wave and particle phenomena together with the measurement of the velocity all depend upon that; there are always 'grids' of one kind or another involved. The temptation to argue backwards from phenomenon to essence is so strong that most regard it as churlish to put forward such nit-picking arguments. However, the evident difficulty imposed by the facts and the extreme sophistication of the current physical theories required to explain them suggest that perhaps such fundamental considerations are justified. A simple mistake in immediate inference is of course embarrassing, but it is worth facing up to.

Our approach is based on the idea that just as there are three fundamental point-based linkages between space and counterspace where space dominates and counterspace absorbs the strain, then there should be three fundamental plane-related linkages where counterspace dominates and space absorbs the strain. The three flavours of this are for *polar affine*, *polar special affine* and *polar metric* linkages, symmetrical to the use of affine, special-affine and metric linkages for the elements. We propose that light is related to the first of these. This assumption has proved fruitful and opened up the way to approach the other two. We need now to understand what we mean by 'polar affine.'

We saw that an affine transformation leaves the plane at infinity invariant. In fact it also transforms straight lines into straight lines and flat planes into flat planes. It does not preserve length, angle, volume or area. An important property is that certain ratios are preserved. For example if two rods are parallel and then affinely moved, their lengths may change and also their orientation in space, but interestingly the ratio of their lengths will not change. So if they start off with lengths 6 and 4, say, their final lengths may be 21 and 14, or 3 and 2, or any other pair of numbers in the ratio 3:2. They will end up parallel, as parallelism is conserved by such transformations. Thus if we have a standard measuring rod we can arrive at the lengths of all other rods parallel to it, and their ratios are conserved in an affine environment. On the other hand if we have two rods that are not parallel then the ratio of their lengths is not conserved. This means that we cannot make comparisons in different directions.

A similar state of affairs exists for areas. If we have two squares in a plane and one has twice the area of the other, then if we transform the plane in an affine manner the final plane will contain two parallelograms instead of squares (as angles and lengths will probably have changed), but one will still have an area twice that of the other. Again this applies if the squares lie in two parallel planes. On the other hand if they lie in different planes that are not parallel then the resulting parallelograms will not necessarily have the ratio 2:1. Thus although area itself is not defined for affine geometry, oddly enough the ratio of parallel areas is. We can interpret that most easily if we think of an affine transformation occurring in Euclidean space, as then we can sneakily use the Euclidean background to measure the areas and thus find that they stay in the same ratio. In the same way we find that ratios of volumes are also conserved.

This all concerns ordinary affine geometry. If we now grit our teeth and polarize the whole thing we can describe what goes on in polar affine geometry. The first step is to realize that it leaves the ideal point at infinity of counterspace unmoved. Projective geometry is identical in both space and counterspace as it is inherently self-dual as we saw earlier on, but as soon as we select something to play the role of infinity we either go towards space by selecting a plane or towards counterspace by selecting a point. In the latter case we must work with ratios of turns instead of ratios of lengths. This is harder to visualize as we cannot take simple rods but must work with pairs of planes. We might choose a wedge of cheese as a picture of a turn, so that counterspace cheese has a turn between its planes instead of an angle. (We hope the taste is invariant, but that is not a geometrical issue!) Now to illustrate the ratio of turn rule we need two wedges of cheese with different turns (see Figure 16). In the spatial case we had two parallel rods, so somehow we must see what it means to say that two wedges of cheese are 'parallel' in counterspace and thus also in a polar affine transformation. Recall that in space two lines are parallel if they meet in an ideal point lying in the plane at infinity. Hence in counterspace two lines are parallel if they 'meet' in an ideal plane 'lying in the point at infinity.' This means that our two wedges of cheese must have their edges (where their planes meet) lying in a plane which contains the ideal point we have selected as invariant (say *O*). In that case when we transform them so as to leave *O* fixed then the two new wedges will have the same ratio of turns as the old. If the edges do not lie in the same plane, or they share a plane that does not contain *O*, then a polar affine transformation will not preserve that ratio intact.

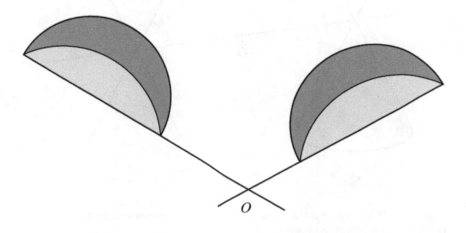

Figure 16. Polar affine transformation.

More important for light is the preservation of the ratio of polar areas. Recall that we gave the example of the polar area of a cone as the equivalent in counterspace of the area of a circle in space. The polar area of a cone is composed of all the planes containing its vertex which do not cut the cone. Now just as an affine transformation preserves the ratio of areas in the same plane, so a polar affine transformation preserves the ratio of polar areas 'in the same point.' For two cones that means they must share the same vertex, which is a neat and simple rule (for a change). However simplicity rarely persists and we must now recall that the ratio of two areas is also preserved if they lie in parallel planes. We remember that in counterspace we have parallel points instead of planes, which now suddenly becomes a helpful idea as we will need it for our two cones. We saw that two points are parallel if they lie on a line containing *O*, so if the vertices of our two cones lie on a line through *O* then a polar affine transformation will leave the ratios of their polar areas fixed. (Figure 17)

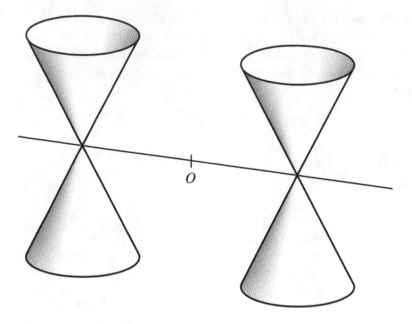

*Figure 17. The ratio of polar areas of two cones is preserved if their
vertices lie on a line through O.*

Light has some particular properties of importance. It is highly
directional. Two beams of light in distinct directions that cross do not
affect each other. We are reminded of affine incomparability in different
directions. Light is also expansive, an affine attribute. It is space filling
and does not in any sense 'preserve volume,' polar or otherwise. It can
neither be seen nor measured unless it interacts with matter, and then as
we have already noted we have to be cautious in the interpretation of
what the measurements actually mean.

For these reasons among others the idea that light may be related to
polar affine transformations seemed worth exploring. But what is the
linkage agent when light does interact? For the elements we simply
used point linkages, but now we have to find something else suitable.
We require something that, like vectors, is independent of the way we
describe the situation. The ratios available are those between turns, or
polar areas, or polar volumes. The latter does not seem to have any connection with the above properties of light, and turning planes do not
either, so polar area was selected. This has two immediate advantages: it
is a two-dimensional element in line with Steiner's finding about ethers,
and if we select the polar area of a cone as our linkage element then it

is describable mathematically as a certain kind of vector, albeit more sophisticated than the ones we have met so far, known as a *bivector*. An example of a bivector is angular momentum. If a circular metal disc is rotating it resists having its axis of rotation changed (the gyroscopic effect). The larger the diameter of the disc for a given rate of rotation, the greater is that effect, which thus depends upon the area of the disc for a given thickness. What resists the change is called angular momentum, which is related to the area of the disc. Its speed of rotation is also important, but for this illustration we assume that is unchanging. This new quantity, angular momentum, behaves like a vector in three dimensions as it is a physical quantity that does not depend upon how it is described geometrically. However the direction involved is not like an ordinary velocity which is described by a line of a given length. Instead it is a planar quantity as we have tried to illustrate. For this reason it is given a special name to make clear the distinction, namely a bivector.

Returning to the cone, it is plane-based as the polar area of a cone is made up of all the planes in its vertex that do not otherwise cut its surface, so we are now looking at a linkage based on planes, although not simply on single planes. (An animated illustration of this may be found on *www.nct.anth.org.uk/ethers.*) In this way the idea arose that what are ordinarily called photons are actually counterspace cones. Now a conventional photon has three basic properties: direction, spin and energy. A counterspace cone as a bivector also has three basic properties: direction, a sense of rotation, and polar area. We imagine the cones existing in counterspace and thus possessing definite polar areas, but acting according to the rules of polar affine transformations. This immediately suggests that there is a relation between polar area and energy. Conventionally, energy is related to colour in the case of light. Thus we expect the polar area to have some connection with colour. However we recall our earlier comments on colour, and prefer to say that polar area is a contingency for the manifestation of colour, not colour itself.

We thus have a candidate for photons that is neither a particle nor a wave, which in line with our earlier discussion may then manifest wave-like or particle-like behaviour according to the interaction involved. It is like the cloth which may appear grid-like but is not itself grid-like. We now need to see how phenomena like reflection, refraction, diffraction, polarization, scatter, emission and absorption arise from this proposal. We will not attempt to explain them all here, and the interested reader can find these in my earlier book (Thomas 1999, Chap. 13). That they can be explained using photon cones is remarkable and shows that our approach is fruitful. However we will give some examples to show the line of reasoning involved. It should however be noted that we are now considering light when linked to space, and thus it has 'taken a step down' from pure etheric light. We content ourselves with this for the moment as our overall aim is to relate the interactions between space and counterspace to observable phenomena.

While we are on the subject of photons we may illustrate the potential explanatory power of our approach in relation to some 'spooky' experiments done with light.

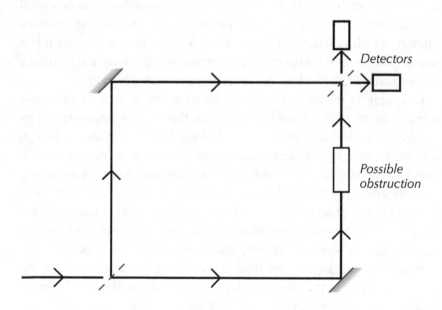

Figure 18. A beam of light split by a semi-silvered mirror.

Figure 18 shows a photon at the bottom left approaching a special mirror which is semi-silvered, which means that there is an even chance it will be reflected upwards in the diagram or carry straight on. There are two other ordinary mirrors so that the photon ends up at another semi-silvered mirror at the top right, either travelling upwards in the diagram or to the right, where again it may be reflected or not with a 50:50 chance either way. The two detectors shown tell us where it ends up. If there is no obstruction in either path, then oddly enough one of the detectors always fires and not the other. In other words, when the experiment is repeated many times it is always the same detector that is activated. However if an obstruction is placed in one path as shown then either detector may fire with equal likelihood. It requires sophisticated quantum physics to explain this extraordinary result (Penrose 1994, p. 267). Like other similar experiments it is as though the photon 'knows in advance' whether or not an obstruction is present. Now consider our concept of the photon. As a cone with its axis along the path of approach to the lower left semi-silvered mirror its polar area will 'embrace' the whole apparatus and thus it is no longer surprising that it 'knows in advance.'

As an example of the use of photon cones (as we shall call them) we will examine diffraction at an edge, but we need some preliminary introduction. When a photon is emitted from an object the photon cone expands until it becomes a cylinder. Now this will be a minutely thin cylinder and since the polar area is all round the cylinder (from a spatial perspective) we see that the light as such is all round the cylinder so that the inside of the photon cylinder amounts to *a ray of darkness*. This suddenly elucidates one of Steiner's observations that light does not consist of 'rays.' Yet the ray model is most explanatory when studying how light passes through lenses and reflects off curved mirrors. It works because it is unsuspectingly operating with rays of darkness!

Returning to the emitted photon cylinder, if there is nothing other than the source CSI (counterspace infinity) in the cylinder then no strain arises. But if an obstruction extends into the cylinder then its CSIs being counterspace infinitudes, cause strain as the polar area of the photon cylinder seems different for each. A single example of this difference may indicate why that is the case. In ordinary space a circle has a finite area, whereas a parabola has an infinite area. That is because it reaches to infinity. The polar situation for a photon cone is to have a CSI on its axis, in which case it is like the circle, having a finite polar area. However a CSI located on its surface causes it, *for that CSI*, to have an infinite polar area because it 'reaches to infinity,' that is, to the CSI. CSIs located between these two extremes 'see' it with different polar areas.

So the photon cylinder considered above realigns itself so that there is no obstruction in it (if possible). To do so it is necessary to be able to compare the polar areas of the old and new photon cones, and we recall that an affine comparison must occur at a point (or at two polar parallel points which is not suitable in this case).

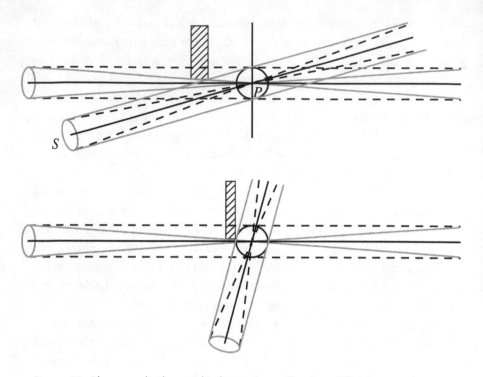

Figure 19. Photon cylinders with obstructions showing diffraction.

The top diagram in Figure 19 shows a photon cylinder with an intruding obstruction and the most 'expanded' possible cone which excludes the object. The vertex is at *P* and is the interaction point for an affine comparison. Another cylinder is shown related to a new cone sharing the vertex *P* and such that it just clears the obstruction. The new cylinder must have the same polar area as the original (or, to be affinely strict, have a ratio to it of 1). The lower diagram shows a similar arrangement but with the obstruction closer to the axis of the cylinder so that the deviation is greater. This captures some of the essential features of diffraction: the obstruction must be sufficiently close to the axis (of the same order of magnitude as a wavelength conventionally, which indicates how wavelength may be re-interpreted), and the closer it is, the greater the deviation. Also, the greater the polar radius the closer the obstruction must be for the same deviation, relating to the difference caused by colour. An observer will see the diffracted photon as if it were emitted from a source *S* located where the new photon cone and photon cylinder intersect. This captures the essential features of the phenomenon.

Reflection is very easily exhibited and the usual law for a flat mirror is obtained where the incident and reflected angles are equal. The turn between planes is invariant across the reflection. We take the base plane for this as the plane α tangential to the surface, and relate all planes in the polar area to it. The planes in the polar area meet α in lines, and the reflected planes forming the polar area of the emergent cone will share those lines and their turns will be conserved. This gives the observed reflection law as shown for two representative planes in Figure 20 below.

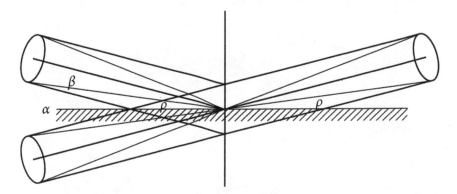

Figure 20. Photon cylinder showing reflection.

The incident cone and cylinder are shown on the left above the mirror, and the turn ρ between α and the tangent plane β is shown, which is conserved through the reflection. The turn ρ of the reflected tangent plane will be measured with respect to a virtual source located where the reflected cylinder and cone intersect as shown, which agrees with the usual appearance of virtual images. The polar area of the incident and reflected cones is comparable at their coincident vertices, which then gives that of the reflected cylinder. It is clear that all incident and reflected planes making up the polar areas satisfy the stated conditions if the axes of the two cones are in the same plane as the line shown at right angles to the mirror. For curved mirrors the above considerations apply for each tangent plane, yielding the known behaviour of such mirrors without difficulty. This simplicity should be compared with the extraordinary lengths gone to conventionally to explain this simple law.

Refraction occurs when light crosses a boundary between two substances of different density, for instance, air and water. If a coin is placed on the bottom of a glass full of water and viewed obliquely from the top the coin will appear to be raised above the bottom of the glass. This is because the direction of propagation is diverted at the surface of the water. The relation between the angles that these directions make with the surface is given by Snell's Law. This can also be derived based on counterspace principles, and hinges on the fact that there is a different scaling between space and counterspace in air and water which causes the deviation (Thomas 1999, page 94). Conventionally this is explained by a change in the speed of light, and in the next chapter we will see the connection between scaling and the supposed velocity of light.

8. Time

This brings us now to a radical idea about time. When a photon cone is emitted from a counterspace infinity (we assume this as CSIs are the active ingredients of matter) we can think of its vertex moving away from the CSI (which lies on its axis). Now suppose there is an obstruction such as a human eye which intercepts the photon cone. Since affine comparisons must be made at the vertex it follows that the interaction with the eye will be at the vertex of the cone. Now the plane which is the polar centre of the cone (which is polar to the centre of a circle) lies in the vertex and is at right angles to the axis (see Figure 21).

The vertex will be at a distance r, say, from the CSI at O while the centre plane will be at a turn τ_I, say, from the plane at infinity of space. But τ_I is proportional to $1/r_I$ so we have $r_I \times \tau_I = c$ where c is some constant for the medium which governs the scaling between space and counterspace. No matter where the eye intercepts the vertex of the cone,

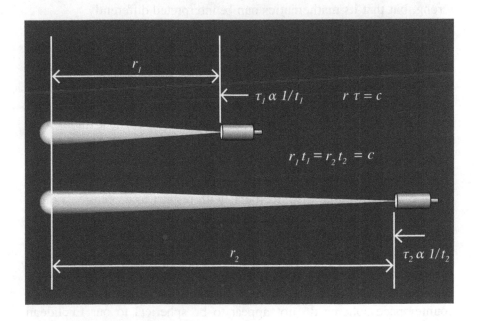

Figure 21. Photon cone emitted from a CSI.

we will have the outward radius multiplied by the turn giving the same result c. Now if we interpret the radial turn τ_1 of the centre plane, that is, its turn inwards from infinity, as the reciprocal of time, then we have $\tau_1 = 1/t$ and our formula becomes $r/t = c$. What this says is that the distance divided by the time of the interaction is a constant. But distance divided by time is usually a *velocity*. Thus although in reality it is a scaling constant, c will *appear* to be a velocity. The light *per se* does not move with any velocity as it exists all round the photon cone, so the term 'the velocity of light' is, from this perspective, a misnomer. This is even clearer when the cone has expanded to a cylinder. Yet we see why light appears to have a velocity. At a stroke we have an explanation for Steiner's finding that light does not have a velocity, and at the same time why in our experiments it will always appear as though it does. But that is not all. The apparent velocity does not depend upon the state of movement of the eye observing it or that of the CSI emitting it. After all it is a scaling constant unaffected by such things, so the law of the constancy of the velocity of light so fundamental to Einstein's Theory of Relativity is explained in quite another way. In any case Einstein did not attempt to explain it, he assumed it and derived his theories accordingly, but we have explained it. We do not need Relativity as we no longer regard light as moving like a physical object, although we are not saying Relativity is wrong, but that its mathematics can be interpreted differently.

This idea about time needs to be considered carefully as it has proved of fundamental importance in subsequent work even if it seems rather like an arbitrary assumption just now. In the course of our research it was far from arbitrary: rather it forced itself upon us! We are saying that the reciprocal of radial turn in counterspace measures time. Not any turn, as was at first assumed some years ago, but just *radial* turn. To clarify the term 'radial' think first of the polar situation in space. If we have a sphere then it has a radius which is the distance of its surface from its centre. If the sphere expands or contracts this radius changes and we think of radial displacements with respect to its centre as those along lines through the centre. Displacements in other directions such as along a tangent are not radial. The polar situation in counterspace is a sphere with the plane at infinity as the polar of the centre of the spatial sphere, that is, a counterspace sphere has a plane as its 'centre' (we only consider the case with the plane at infinity at the centre because other counterspace spheres do not appear to be spherical to our Euclidean consciousness, and are better avoided just now). Any plane has a turn from the plane at infinity that is the polar of the radial distance of a point from the centre in space. Thus radial turns are either between planes and the plane at infinity, or between 'parallel' planes, that is, planes which are Euclideanly parallel, intersecting in a line in the Euclidean plane at infinity. Although there is no parallelism between planes in counterspace we can loosely use the term for convenience. Even that is only possible if the plane at infinity is a linked plane, which we assume is the case to give a basis for time measurements. If two planes are tangential to a sphere centred on a CSI then they clearly have the same turn in from the plane at infinity and thus there is no time interval between them, and if those planes move so as to remain tangential (Figure 22) then no time intervals are involved.

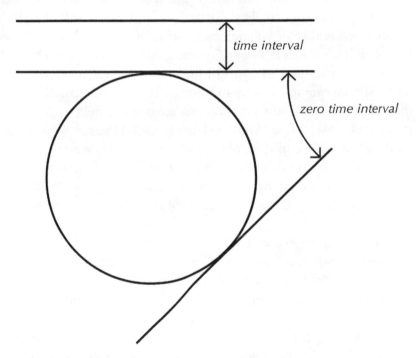

Figure 22. Planes tangential to a sphere centred on a CSI.

Our investigations found that when potential energy in relation to gravitation is interpreted in counterspace terms the requirement arises quite independently that time be related to the reciprocal of radial turn. We will not treat that here (see Thomas 1999, p. 76) but it is mentioned to show that this approach to time makes sense more widely. A striking further corroboration came more recently, namely that when space and counterspace are linked using special equations to satisfy certain technical requirements of physics, the spherical wave equation is obtained as a result. This is the equation that describes the behaviour of waves such as sound waves or explosions that expand outwards on a spherical wavefront. The application to counterspace is however quite different (it arose during the search for a detailed understanding of the absorption of a photon cone) and is the subject of ongoing research.

We come now to our fundamental thesis about the ethers. Steiner referred to four ethers: the light, chemical, warmth and life ethers. We have explored light and found a basis for it which now suggests a new approach. Remembering his finding that ether is essentially two-dimensional, and noting that time relates to a single dimensional aspect of counterspace — namely the radial — *we propose that the residual two-dimensionality of counterspace (when that is excluded) concerns the ether*. This is subtle as our Euclidean consciousness can all too easily see dimensionality in terms of lines at right angles in space. We can take a spatial example to help us, though. The three dimensions of space may be described using three co-ordinate axes at right angles such as forwards/backwards, left/right and up/down. But there is another way as used on the surface of the Earth where we use latitude and longitude to specify our position and ignore radius while we remain on the surface. We are doing something analogous in counterspace when we ignore radial turn and say there is a remaining two-dimensionality pertaining to the ethers.

Another way of expressing this is to say that *an ether concerns time-invariant transformations in counterspace*. The above line of reasoning most readily yields the idea of action within the surface of a sphere, and we will come to the ether associated with that in a moment. But how does this conform with our study of light? It would seem that radial aspects have essentially been involved there, but there is an important time-invariance involved. A photon cone has a polar area that is invariant with respect to radius and hence with respect to time, and the application of the ideas to explain phenomena all rely on this invariance. It is a two-dimensional aspect of counterspace that is invariant for photons. The expansion of a photon cone to a cylinder after emission can no longer be regarded as a movement of the vertex that takes time, as that would require yet another standard of time, and we prefer to think of it as instantaneous so that we have an ensemble of polar-parallel photon cones all of the same polar area, including the cylinder. Which of the ensemble 'actualizes' and interacts with an obstacle is determined by the conditions (Figure 23), and we saw different cases for a total obstacle like an eye and an intrusion in the cylinder in the case of diffraction. It is a short step to see that emission and absorption are a total process, even a mutual interaction. We see that the time-invariance approach to ether is more serviceable for light, but the first definition elucidates the dimensionality aspects.

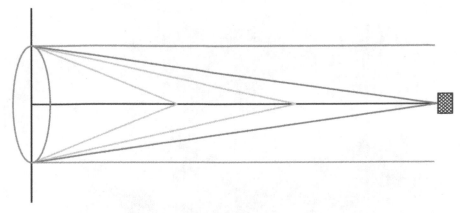

Figure 23. A photon cone expanding to a cylinder.

One point about light that we must mention before we go on concerns extensive versus intensive scaling. The argument about the constant apparent velocity of light has an extensive flavour to make it easier to grasp the basic idea. However more detailed work on how the scaling should be expressed mathematically shows that an intensive non-back-to-back scaling scheme is possible that nevertheless gives the same results.

By *extensive scaling* we mean a scaling relating absolute quantities in space and counterspace, for instance, shift and distance or turn and angle. By *intensive scaling* we mean the relationship between local rates of change, for instance, at a point or in a plane. Thus the rate of change of shift or turn can be scaled locally to the rate of change of length or angle. It may vary across a volume or area! Indeed our use of vectors shows quite clearly that intensive scaling is involved.

Having laid the basis for an approach to the ethers and applied that to light, we will now move on to the other ethers.

9. Chemical and Life Ethers

Chemical ether

We saw in the previous chapter that we have time-invariance for processes in a spherical surface in counterspace, and we expect this to relate to a different kind of ether. We need to find in what ways turns of planes may vary on a spherical surface to fulfil several conditions that enable that activity to be related to the physical world:

1. There must only be one value of turn at each location of the sphere. In the physical world the height of a point on a hillside, for example, can only have one value. This requirement need not apply to pure etheric action, but must do so when it is linked to the physical.
2. No value must be infinite, which is obvious as we do not have infinitely large physical values.
3. There must be continuity. A mountainside is continuously connected, for example. Its surface can never have a region like a hole in a piece of paper, and thus if we are given any two points in the surface it is always possible to draw a line in the surface from the one point to the other.

These conditions can be met mathematically and the resulting surface variations are interesting. We find that round the sphere we have wave-like variations as the 'longitude' varies such that there must be a whole number of wavelengths for one complete circuit, the same at all latitudes. This introduces whole numbers and waves. As the latitude varies we have definite types of possible variations changing rhythmically with longitude, and a typical picture of this is shown in Figure 24.

These kinds of distribution occur in many physical circumstances such as vibrating rings, bouncing balls and tidal motion. Their particular qualities of interest to us are that whole numbers occur and waves are involved. This strongly suggests that the ether described is the *chemical*

Figure 24. A sphere with wave-like variations around it (see inside back cover for colour version).

ether which Steiner also described as the *number ether* and the *tone ether*. We thus find why this ether has these different designations. The polar volume of the counterspace sphere remains constant as that is the condition for its surface to remain time-invariant, so we see this as related to a polar special affine linkage. The mathematics involved is identical to that required for quantum theory so we find that we can relate our work to quantum theory via the chemical ether. It is interesting that time invariance is introduced in quantum physics somewhat arbitrarily to enable the mathematical equations to be solved, whereas we start with it as a definite requirement. That we can make contact with the intricacies of the quantum theory of physics at such an early stage in our enquiries is encouraging. We have used a sphere, but other surfaces such as spheroids may also be considered.

The essence of chemistry can be seen as the study of the way substances react together, for example, water is a compound of hydrogen and oxygen. Reaction is much more intimate than physical processes which merely change shape, size, state or temperature, and so on. When such an intimate chemical bonding occurs the properties of the constituent elements are usually unrecognizable, so that in water with its wetness and liquidity we cannot discern the properties of hydrogen which is the lightest of all elements, is a gas, and is highly explosive. Neither can we recognize the oxygen in water which is also a gas and is required for most burning to take place as well as being essential to us for life. Chemical bonds are thus fundamental to the subject and come in several varieties, one of which is called the *covalent bond* which is a strong bond involved in the binding together of elements in compounds. We will now see how this arises from the above considerations about counterspace.

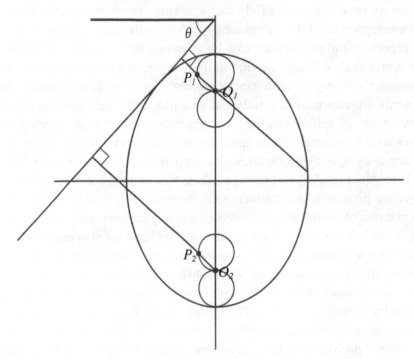

Figure 25. A prolate spheroid with two double-spheres functions showing no strain. An image of chemical bonding.

Figure 25 shows an ellipse representing a prolate spheroid (a rugby-ball shape), with two double-sphere functions at the foci O_1 and O_2, and a tangent to the ellipse. The lines at right angles to the tangent meet the two functions in the points P_1 and P_2 such that $O_1P_1 = O_2P_2$, which shows that the turn of the tangent is the same as seen from both foci. It is possible to describe a bond between two CSIs in this way which points the way to its application to chemical bonds, particularly the covalent bond. Briefly, two CSIs can exist one at each focus of the prolate spheroid and when they do so there is no strain as each sees the surface activity in counterspace as being identical ($O_1P_1 = O_2P_2$), but if either tries to move then that is no longer the case and strain arises. This further corroborates the idea that time-invariant activity in counterspace surfaces relates to the chemical ether, and in particular to chemical bonding.

Life ether

Finally we need to consider fully metric linkages to counterspace where it is counterspace itself that dominates, which we will relate to the life ether. We expect spatial phenomena to arise reflecting the counterspatial dominance, and as an example we find in many organisms the phenomenon of *phyllotaxis* where growth relates to spacing. For example in a sunflower the seeds are arranged in spirals and on a pineapple there are two interweaving sets of spirals, but the term originally referred to the spacing of leaves on a stem where again spiralling often occurs up the stem and the spacing obeys a mathematical law. It is particularly noteworthy that we do not have equal spacing and George Adams first pointed out the relation to projective geometry and counterspace (Adams 1980, p. 213). We see all this in terms of spatial form following counterspace laws since they dominate in the linkage. The linkage agent being explored is rather complicated mathematically, but it points to such formations because it involves a permanently-acting rotation or 'spin' between space and counterspace and thus quite naturally gives rise to *path curves* which have been extensively studied by Lawrence Edwards (1993). We see such curves in many spiral formations in nature, archetypally in a fir cone. They produce forms with egg-shaped profiles found in such diverse objects as flower and leaf buds, birds' eggs and the heart.

    For these reasons we see this form of linkage as related to life ether. Now a metric linkage is quite rigid and thus seems rather unsuitable for life with its inherent properties of growth and mobility! However the processes required in living organisms are precise and measurable, such as the generation of heat and energy and the precision of the action of enzymes. The aspect of form need not only be thought of in terms of rigid structures such as we find in crystals in the case of mineral linkages, and in particular our attention is drawn to surface structures which remain more plastic and yet involve important quantitative processes. Membranes are such surfaces, and they occur extensively in organisms. The skin is a membrane, all our internal organs such as the liver are surrounded by membranes, and every cell of the body is enclosed in a plasma membrane. The importance of membranes lies in their filtering ability, for they control what substances may cross them and in which direction. The plasma membrane ensures that the chemical balance and liquid content of a cell satisfy its organic needs by controlling the diffusion of substances like potassium and sodium into and out of the cell. Membranes are usually *semi-permeable* which means that some substances, including water, can cross them one way but not the other. Membranes in the brain keep out poisons, and if damaged by drugs result in the incursion of substances that lead to hallucinations and the like.

Geometrically a membrane has an inside and an outside. Physically this is obvious, but if a membrane is linked to both space and counterspace then from a counterspace perspective those terms are reversed, and the 'outside' is for counterspace what we would normally call the inside and *vice versa*. If we apply this idea to a cell then all other cells of the body are 'inside' it for counterspace. Moreover this is true for every cell of the body and we begin to see *what makes a living body an organism: it is inside every cell for counterspace so that each cell relates to the whole organism just for that reason.* The images of the organism seen by every cell must differ for a many-celled body and yet they need to be in synergy. We have a complex counterspace structure which acts like an archetype for the organism, manifesting in its overall form and structure so that strain or imbalance of the synergy caused by illness or injury result in healing processes that attempt to restore it.

The spindle tree shown in Figure 26 thus lies inside the cell indicated for counterspace.

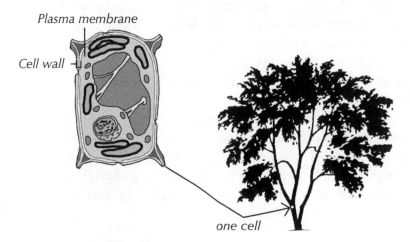

Figure 26. The cell inside a tree. In counterspace the tree is 'inside' the cell.

We should at this point mention the contrast between the view espoused here and *vitalism*. The latter supposes that the laws of physics and chemistry are insufficient to account for life and consequently proposes an agent that changes those laws inside living organisms. However the extensive research that has been done on organic processes has certainly revealed an almost bewildering capacity for organization to arise, but that does not seem to require any changes in the laws of physics and chemistry. Rather the reverse, in fact: those laws seem to be just right for organisms! It is not that they explain organic processes so much as that they provide just the right foundation upon which to build. The builder remains rather obscure and mainstream science would prefer to do without one, relying on the 'blind watchmaker' of gene-driven Darwinism.

So vitalism does not seem to be the right approach, and yet the wonderful capacity for self-organization and the kinds of law exhibited by organisms seem to point to something quite distinctive which reductionism loses rather than explains. The concept of the life ether is not that of a vitalist principle but rather that very agency which gives substances their properties to support life, and contains time-like structures which unfold into spatial ones. We have in mind structures in counterspace that in themselves are complex planar arrangements with definite turns and shifts, but which are not directly linked to space in an extensive sense. The linkage is via time, so that what exists in counterspace gives rise to processes in space that use the laws of physics and chemistry much as a painter uses paints, but just as reduction of a painting to its constituent colours loses the painting, so reduction to those laws alone loses the organism. This stares us in the face and yet it is so easily, even wilfully, overlooked. The concepts associated with life are not readily dissolved by plunging into genetic explanations as the very genes themselves end up having organic properties such as self-repair, making it plain that nothing fundamental has been achieved by that reduction even if much remarkable detail about how substances are orchestrated into organic action has been exhibited.

Our aim is to take hold of life processes in their own terms and the counterspace approach seems promising. Adams pioneered this and his books and those of Olive Whicher (Adams & Whicher 1980) can be consulted to see the 'gesture' of the idea, but we aim to take hold of it in the same detail as we have begun to accomplish in other areas, even if that aim remains for the future. The approach sketched here could point the way to that.

10. Science and Counterspace

We can now sum up the position we have reached in seeing various aspects of science in relation to linkages between space and counterspace:

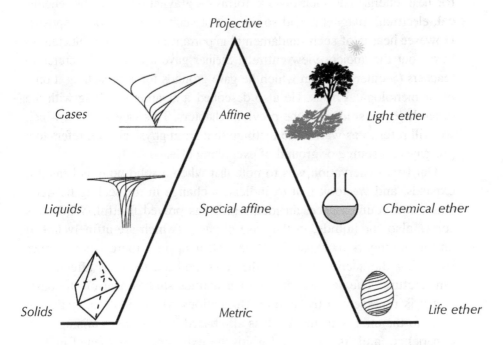

Figure 27. Special geometries and physics.

At the top of Figure 27, we have projective geometry of which the geometries lower down are special cases. On the left we show the different pointwise linkages related to the gases, liquids and solids, and on the right the planewise linkages of polar area and volume related to light and chemistry, and finally the metric linkage for life itself. This constitutes a fairly comprehensive approach to science and is a kind of map of our research programme. In each area we have attempted to show how it may be approached and have derived some quantitative as well as qualitative results. Both are important as to concentrate on the one at the expense of the other would risk an imbalance in our approach.

So far we have only treated heat in passing. It is not explicitly shown above but in a sense pervades the whole diagram, which does rather accord with its nature. As we pointed out before we cannot simply place it at the top with projective geometry as the latter is strictly non-metric and so does not afford any measurement, and yet heat enters into quantitative relationships of various kinds. While the ancients regarded it as a separate state of matter or 'element' (fire), modern thought sees it only as a modification of matter due to energy. While it might be argued that energy is another name for heat, energy has such diverse forms — gravitational, kinetic, chemical, electrical, magnetic, and so on — that such a view is too simplistic. However heat is of such fundamental importance that we are reluctant to throw out the ancient view entirely. Steiner gave a course of lectures to teachers (Steiner 1920) in which he gave grounds for approaching it on a phenomenological basis. He also described it as an ether along with the other three we studied in the previous chapters. This view of heat, which we will refer to as warmth to distinguish it from physical heat, refers to a pre-physical source or ground of everything else.

Our first observation was to note that when a solid body is heated it expands, and we took that to indicate a change in the scaling between space and counterspace, an idea which has proved fruitful. While this serves also for liquids, in the case of gases (which are affinely linked and no strangers to expansion) we took it up in a more strict manner by seeing the scaling between the strain and the stress as affected by temperature. The science of thermodynamics studies heat on the basis that it is represented by random fluctuations which relate in statistical ways to the energy of matter. It is also based on postulates arising from experience, and its first law forbids the existence of perpetual motion machines of the first kind while the second forbids those of the second kind. An example of the first is an overbalanced wheel such as the eighteenth century inventor Johann Bessler claimed to have made (Collins 1997) while the second kind would convert heat into mechanical work without wasting any heat in the process. The second law also forbids heat to flow from a higher temperature to a lower temperature source, so that if you place your hands in cold water the heat flows from your hands to the water and not *vice versa*. This prevents us from garnering much energy from waste heat and restricts the efficiency of car engines to about 30%.

We regard any source of work as based on strain between space and counterspace so that heat as a source of work may be expected to arise similarly. We have not studied a hybrid linkage where space and counterspace share the stress, and in view of its dual nature as warmth and heat we might expect such a linkage here. A suitable scaling scheme has been found which has the interesting property that it is not reversible: it only scales for changes in one direction. This accords well with heat and its irreversible processes. By this we mean, for example, that if heat is converted into mechanical work then some heat is always wasted in the process such that we cannot perform exactly the reverse process of generating heat from work as the wasted heat cannot be included. Now the scaling between space and counterspace can change locally as is clear since objects can have different temperatures, or themselves suffer temperature gradients, and this will cause strain in relation to the primal counterspace involved, namely *scaling strain*. We see this as the source of work done by heat, and its random properties will explain why statistical methods are needed in thermodynamics. The unidirectional scaling will explain why in the normal course of events perpetual motion machines are not possible, but it should be noted that there is another scaling scheme for a reverse scaling which, when combined with the first in alternation, holds out the possibility of etheric warmth/heat cycles that will appear to be perpetual motion from a physical perspective, although in reality they are not so, as they draw upon warmth (not heat).

One of the most obvious properties of heat is the phenomenon of conduction where a warm body in contact with a cooler one transfers heat to the latter. Also if a body such as a metal bar is heated up at one end then the temperature of the rest of the bar rises as though the heat travels or flows through the bar from the hot end. We imagine each CSI of a substance to be scaled individually and different CSIs are related to one another via the primal counterspace involved so that strain arises because they relate differently to it if their scalings differ, causing strain. This indirect relation between CSIs we refer to as a *fractal linkage* because it applies simultaneously for all CSIs regardless of the size of the objects or separation of the CSIs involved. As a picture of what we have in mind, imagine a large metal ring, suspended from the ceiling, from which a number of smaller rings are suspended each on its own thread. The rings affect each other via the large ring rather than through direct contact, so that if their weights differ an overall balance arises between them via the main ring. If a ring is pulled down all the other rings will be affected. We see the main ring as an analogy of the primal counterspace and the smaller ones as analogous to CSIs. Just as imbalances between the rings cause the main ring to tilt and thus affect other rings, so heat conduction occurs between CSIs via the primal counterspace. This picture differs from the conventional view which regards conduction as arising from the mutual 'jostling' of neighbouring particles. It throws light on Steiner's interesting picture (Steiner 1920) of a number of street urchins under the heated bar who individually cry out in sequence, the point being that the cry of one is not caused by the cry of its neighbour but happens individually due to an underlying cause. Again when we think our premises through, we find light thrown on Steiner's research, often when we least expect it! The relation to time is subtle here as the fractal linkage might be supposed to result in an instantaneous change in temperature of all CSIs at once. What happens is that counterspace links back to space via time such that the greater the spatial separation of two CSIs the greater the apparent time between them, so that a change in a neighbouring CSI will manifest in space sooner than in a more distant one even though for counterspace the two changes are 'instantaneous.' This further elucidates the 'urchin' picture. We must of course bear in mind that our ring analogy is somewhat limited as we are actually concerned with a three-dimensional situation.

To sum up, temperature is proportional to the local scaling between space and counterspace and heat appears as the stress caused by scaling strain. Conduction occurs via the fractal linkage. The scaling involved cannot apply to gravity which shows no discernible dependence on heat.

11. Astronomy and Cosmology

What is our universe really like? The old Ptolemaic system placed the Earth at the centre in accordance with the obvious 'appearances' that it feels to us as if it is motionless and the Sun and stars appear to go round it. Aristarchus of Samos was accused of betraying the Eleusinian Mysteries of Ancient Greece by teaching that the Earth goes round the Sun, although historically this fact has been recorded as his having been indicted for impiety (*Encylopaedia Britannica*). But the Greek idea that circles are perfect, that beyond the sphere of the Moon all must be perfect and therefore changeless, leading to the notion that the planets must travel on circular orbits, has fallen into disrepute.

Copernicus placed the Sun at the centre with the planets on circular orbits, and then Kepler showed that the orbits are more generally ellipses. The cosmos is supposed to be very large and expanding. This does not mean merely that the outermost periphery is being extended by the outward movement of galaxies, but that space itself is expanding, rather like the surface of an inflating balloon with ink dots on it. Our three-dimensional space is supposed to be roughly analogous to the two-dimensional surface of the balloon.

Cosmic implications of counterspace

How big, then, is our universe? What kind of law describes it? One such law is the Copernican cosmology, which is distinct from the Copernican description of the solar system (Rudnicki 1995, Chap 4). It says that the universe as observed from any position looks broadly the same. A generalization of this says that every position in the universe may be taken as its centre, and that its broad features will appear similar in every direction for any chosen centre. This discounts very close objects such as planets in a solar system, paying attention more to the distribution of galaxies. The 'perfect cosmological principle' adds to this the idea that it is true at all times. One might contrast with this an evolutionary law such as that espoused by the Big Bang theory, and the resulting question as to whether there will eventually be a 'big crunch,' or whether expansion will go on forever.

The scientific investigation of our cosmos, based on purely physical assumptions and data, tries to answer this question by basing itself on the General Theory of Relativity together with an interpretation of the red shift of stars as being related to their distance from us, together with various models of stars and galaxies. The use of Cepheid variable stars is important as they systematically vary in brightness with a period strictly related to their luminosity. From that period the absolute luminosity may be deduced, and comparison of that with the apparent luminosity actually observed enables the distance of the star to be calculated. Such a star is called a *standard candle*. The most fundamental and pervasive assumption underlying all this is that light has a constant velocity *in vacuo*, which we have seen is not supported by our investigation into counterspace (see Chapter 8). We interpret what is conventionally called a velocity as being instead a scaling constant between the two spaces, and we saw that the physical evidence supporting Relativity may be accounted for quite differently. Obviously this has profound significance for cosmology.

Modern cosmology involves Hubble's proposal that the universe is expanding, and there is a constant (the Hubble constant) relating the distance of a galaxy from any observer to the velocity of recession of that galaxy, interpreting the red shift as a Doppler shift. The latter is a change in perceived frequency caused by relative movement between a source of sound or radiation and an observer. The Hubble constant, first discovered and roughly measured by Minkowski and Wilson, is a ratio with the dimension of the reciprocal of time, that is, velocity divided by distance. This is indeed interesting when compared with the proposal that in counterspace radial displacement from a *counterspace infinitude* is also proportional to the reciprocal of time. The fact that the Hubble constant does not explicitly — but only implicitly — involve distance is thus significant, noting that distance is a property of physical space but not of counterspace.

If the observed phenomena do indeed involve counterspace then the reason that this parameter is constant may precisely be that it originates in an interaction between space and counterspace, the apparent relation to distance being a spurious interpretation which naturally leads to the 'cancellation' of distance in the ratio. It has long been noted that no phenomena other than the red shift suggest the enormous velocities entailed by a one-sided physical interpretation, and indeed strenuous efforts were made in the last century to find alternative explanations for the red shift. They failed as they were understandably based upon purely physical considerations, whereas the inclusion of counterspace brings in other possibilities. It is suggestive that all our knowledge of cosmology is based solely upon radiation phenomena, in particular light, and light exists primarily in counterspace and only manifests physically through photon-cone linkages. It is reasonable to suppose that such linkages are of local origin, in line with some of Steiner's suggestions about the interpretation of Relativity. Furthermore its wavelike or particle-like manifestations only arise when it interacts with matter, the one or the other according to the nature of that interaction. This would explain the failure of attempts to account for the red shift based on the interaction of light with gas existing between the observer and the galaxy involved, as that would imply premature (far distant) linkages between space and counterspace in addition to local ones.

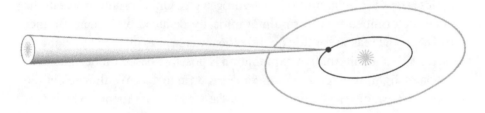

Figure 28. A photon cone relating a star to the solar system.

Figure 28 shows a photon cone relating a distant star to the solar system. Nothing has 'travelled' between the two, as emission and absorption are coincident in counterspace but appear separated in time in space. The photon cone can only link to or react with CSIs, and does not actualize until it does so. Hence the emission occurs only if a 'target' CSI in some other solar system is suitably placed. This eliminates any time of travel and hence all deductions based on it. The process also indicates why even very faint far off galaxies shine steadily rather than flicking on and off stochastically when one of its photons happens to come our way. There are no 'loose photons' that may travel for ever unmolested. The only ones are those mediating between a source and sink. We even challenge the assumption that we are in the same space as other stars. We only think so because we assume light travels to us through space from those stars. But that is not a necessary assumption. We see the stars because of a purely etheric or counterspace connection to them which allows photon cones to be established between us and them. Thus Alpha Centaura, for example, may not share any space with us, so that a space ship launched from us could not arrive there as no space connects us.

The proposal here, then, is that physicality is local and the cosmos is much smaller than is customarily supposed, which also accords with Steiner's spiritual research. It has always been difficult to reconcile his research with mainstream findings, but now it may be possible to make an approach to this with the help of counterspace.

We saw that time is related to the radial dimension of counterspace (see Chapter 8), and if the universe has a definite age then there is a furthest possible plane outwards from us which is close to the plane at infinity. The turn between the two is very small and may lie behind some fundamental constants of physics such as Planck's constant, and on this basis we may account for the Balmer and other related series (cf. French and Taylor 1979, p. 15, for example), recalling that the Balmer series arises for light emitted by hydrogen gas. Only certain wavelengths occur determined by a formula empirically deduced by Johann Balmer in 1885. The reason for this quantization was later explained by quantum physics, but an alternative explanation is possible based on our interpretation of light, provided the universe has a finite age. We thus reject the 'perfect cosmological principle' and the steady-state theory which was based upon it.

We provisionally summarize this as follows:

— physicality is local to the solar system, or perhaps our galaxy;
— other galaxies are in distinct physical spaces, that is, not in ours;
— our only connection with them is via (etheric) light and other radiation;
— such light interacts with matter in our local space to produce photon cones, which are misinterpreted as having physically originated far away;
— the fact that light does not have a finite velocity removes the main basis for interpreting the red shift as indicative of a Doppler shift;
— the property that time is the reciprocal of radial displace ment in counterspace (which is not a distance) gives rise to the Hubble constant which is then misinterpreted as a ratio of

velocity over distance due to the supposition that light has a finite velocity;

— the origin of the red shift possibly lies in the fact that the scaling constant for light in other galaxies differs from that in our own.

Advantages of this approach are:

1. The absurd magnitudes of mass and energy proposed by astrophysics in connection with black holes, neutron star density, and so on, are removed. These have only ceased to seem absurd through usage.

2. The concept of a black hole with its implausible singularity is a materialistic distortion of the idea of counterspace which has a genuine non-physical singularity. The supposed physical singularity of a black hole is thus unnecessary, together with the philosophically unsolved concept of what happens to matter there. Noteworthy is the notion from Relativity that time is different inside a black hole, tending towards infinity as the singularity is approached. But this is just how time behaves in counterspace. It has recently been admitted by some astronomers that the theory of black holes is not as conclusive as is generally indicated to the public (*New Scientist,* January 19, 2002).

The mass of the Sun

The mass of the Sun is assumed to be enormous, for instance, using Newton's law of gravitation to calculate it. But gravity may be accounted for as a form of stress between space and counterspace. It does not require both gravitating objects to be massive, so the Sun may be regarded as a 'strong' counterspace with photon cone linkages to space occurring at its surface giving rise to the coronal and other phenomena actually observed. The 'strength' we refer to is that of the linkage between the two spaces which — being a linkage — involves a non-physical component, making contact with Steiner's extraordinary conclusion that we would find 'less than nothing' if we could somehow penetrate inside the Sun. If the Sun is essentially a 'strong counterspace' rather than a nuclear fusion reactor as currently envisaged scientifically, then that finding could be understood.

Then the Sun's apparently material nature at the surface could be envisaged as a kind of 'tearing' process of the Sun at a frontier between space and the counterspace, resulting in CSIs which form the many gases present. All knowledge of the interior of the Sun is necessarily obtained indirectly through material observations, and when the influence of etheric processes impacts on events in space it is possible to re-interpret them in physical terms which may be logical and consistent, but yet may not be the truth. A very simple example of this is when we suck a drink through a straw. We are conscious of suction and know what we mean by it, but suction can easily be reinterpreted physically as arising from the effect of a difference in pressure between the atmosphere and our lungs. The point turns on how a process is initiated, from space or from counterspace. In the case of suction a counterspace process initiates the action, but of course it results in changes in physical pressure. Similarly the true processes occurring in the Sun may well be etheric in that the linkages are planar rather than pointwise, but our ordinary consciousness re-interprets the results as (increasingly peculiar) physical processes. The concept of the neutrino problem may perhaps serve as an example.

A neutrino is a hypothetical particle in the sense that it cannot be observed directly but is postulated to exist to 'balance the books' for various conservation laws. It was not supposed to have any rest-mass, which means that were it brought to rest it would vanish! Efforts to resolve a discrepancy between the theoretical and observed neutrino flux from the Sun finally led to the conclusion the neutrinos do after all have rest mass. Our point is that the physical entities postulated to explain what is observed are no less bizarre than our proposals about counterspace and ether and arise — we suggest — because etheric causes are being reinterpreted physically.

We come now to a fundamental question that lies at the heart of our whole approach to counterspace and bears on what can now be said about the Sun and stars. We have spoken about linkages between space and counterspace in rather abstract terms, assuming 'stitches,' vectors, bivectors and so forth. But what actually establishes and maintains such a linkage? Our own activity in exploring this is that of thinking, and at the outset we explained that when the word 'spiritual' is used only that activity should be assumed, not as a brain process but as a non-sense-perceptible non-physical process (see Chapter 1). Since thinking serves us well in the sense-perceptible world it is clearly able to bridge between that world and the non-physical dimensions of existence entailed in the notion of counterspace. *We thus propose that it is some kind of thinking activity that lies behind linkages.* This may seem a big step, but without it we remain in abstractions. We will make use of the term 'being' for convenience and brevity, but only in the special sense that we refer to the bearer of the thinking activity behind the linkages. Of course this requires us to suppose that a thinking activity may exist other than in human beings. We do not blush because it is reasonable, more so than some of the concepts blithely accepted in science today, for instance, the unbelievably enormous energies required in string theory or in the theory of black holes, not to mention 'dark energy' and 'dark matter' in modern cosmology which have no direct experimental basis or con-firmation. Thinking, however, does have that — through our very own direct experience.

We may then suppose that beings in this sense are involved in sus-taining the strong linkage between space and counterspace that makes the Sun, and if we turn to stars then we may imagine something similar occurring there, and stellar evolution is an impressive picture of their life-cycle, the linkages waning or changing type as the stars get old. The cataclysmic explosion associated with a nova or a supernova points to a severe crisis at the end of the life-cycle. We recall that metric geometry is not what appears to our direct perception, but rather affine geometry. We think metric geometry as a result of our experience, and if that geometry is actually creatively thought by the powerful thinking activity of beings, then as they withdraw from manifestation the geometry changes and then collapses, as witnessed by the supernova and possible subsequent black hole, although we have reservations about black holes as already explained.

Is the Sun an enormous nuclear fusion reactor as conventionally sup-posed? Spiritually this is an important question because the Sun is our common spiritual heart, and to think of that in such terms is disastrous for social life, which is perhaps all too clear even if this is only one of many factors involved. We have seen that it is not necessary to regard the Sun as massive, and a re-think of the source of light and energy it gives us is possible. In particular the idea that the gases observable by us arise from a 'tearing' process between space and counterspace could well explain the energy output, especially when the enormity of its sur-face area is considered. Of course the Sun has no well-defined surface, but the gist of that statement should be clear. Steiner spoke of lightning as arising from a 'tearing' of space, not from electricity. Recent research has been described showing that lightning cannot be accounted for without some 'external' agent, and currently cosmic rays are thought to be that agency (*New Scientist,* April 30, 2005). We are not suggesting that the surface of the Sun is composed of lightning; but considering the power available from lightning, we do suggest that a 'tearing' process in the surface of the Sun could produce enormous amounts of energy.

Planetary orbits

Rudolf Steiner's spiritual research indicated that there are large etheric 'spheres' centred on the Sun and that the physical planets are like pointers delineating their outer radii. Spiritually the impression of 'spheres' arises again and again in Steiner's work, and it is possible now to understand why that might have been.

A spatial prolate spheroid, which is a surface like a rugby ball, may be a 'sphere' in counterspace if a CSI is located at one focus, as we will now see. Just as an ellipse has two foci, and in the case of planetary orbits the Sun is at one focus, so a prolate spheroid also has two foci. This can easily be understood if we imagine an ellipse rotated about its major (that is, longer) axis, for the resulting surface is such a spheroid, and every cross section through that axis has the two foci of the original ellipse. Thus the surface itself has two foci. The case is quite different for an oblate spheroid, which arises if the ellipse is rotated about its minor or shorter axis, yielding a pumpkin shape, as then the foci of the ellipse describe a circle as it is rotated which gives a focal ring. Recall that every point has a polar plane with respect to a quadric surface such as a spheroid, and the polar planes of the foci are especially interesting in counterspace. In the case of the prolate spheroid the polar plane of a focus is outside it, and the smaller the eccentricity the further away it is until in the case of a sphere with zero eccentricity it is at infinity.

If we calculate the turns relative to the polar plane of a focus for all the tangent planes to a prolate spheroid, then when it has its own counterspace infinity at a focus those turns are all the same (Thomas 1999, Appendix 6). But that makes it a sphere in counterspace, for those turns are polar to the distances from the centre of points on a sphere, which are of course all the same. This remarkable fact enables us to throw light on Steiner's reports, for if we take any plane through the focus of the spheroid it will cut it in an ellipse. So if we place the Sun at the focus and imagine the physical planet going round that ellipse then we see how the physical orbits are as observed, while etherically we are concerned with spheres. We may imagine that the plane of the orbit contains the major axis in which case the eccentricity of the orbit is that of the spheroid, or more interestingly we may suppose the axis of the spheroid lies in the Ecliptic in which case the eccentricity of the orbit will be less.

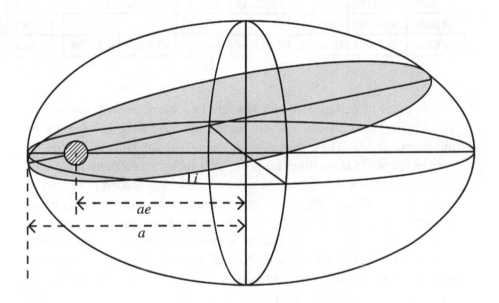

Figure 29. The orbits of planets.

In Figure 29, if we let *a, e* and *i* be the semi-major axis, eccentricity and inclination of an orbit respectively, and suppose the spheroid has its axis in the Ecliptic then we may calculate the spatial semi-major axis *A* of the spheroid, its eccentricity *E* and its polar radius ρ. The distance Z_p of its polar centre from the Sun is the distance of the point where the plane polar to the Sun cuts the axis. The table below shows the results:

Planet	A_{AU}	E	ρ	Z_p	a_{AU}	e	$i_\circ$
Mercury	0.38738	0.207160	2.697358	1.80	0.38710	0.205614	7.00288
Venus	0.72333	0.006833	1.382556	106.04	0.72333	0.006821	3.39363
Earth	1.00000	0.016751	1.000280	59.68	1.00000	0.016751	0.00000
Mars	1.52371	0.093362	0.662066	16.19	1.52369	0.093313	1.85033
Jupiter	5.20281	0.048464	0.192656	107.13	5.20280	0.048452	1.30481
Saturn	9.53980	0.055700	0.105160	170.89	9.53884	0.055647	2.48947
Uranus	19.18189	0.047241	0.052249	405.17	19.18188	0.047237	0.77311
Neptune	30.05790	0.008587	0.033272	3502.01	30.05790	0.008582	1.77283
Pluto	39.58574	0.264931	0.027169	145.42	39.32672	0.253110	17.18036

The units of distance shown here are in Astronomical Units, where 1 AU is the mean distance of the Earth from the Sun. ρ is a turn given in reciprocal AU. These values were calculated for January 1900. The various numbers describing an orbit, such as its eccentricity and inclination, are called the *orbital elements*, and six are needed to specify it completely. Because the orbits are constantly changing slightly, mean elements were used for all the planets except Pluto for which no such elements have yet been derived, as its period of rotation round the Sun is 248 years, but it was only discovered in 1933! So-called *osculating elements* are used in that case to give a feel for it. These are the instantaneous orbital elements at the time of interest, which can really only be relied upon to determine the position and velocity at that time, but not to make accurate predictions beyond about forty days. After that fresh osculating elements must be calculated.

Thus the only planets with their centre-planes (cf. Z_p) inside the solar system are Mercury and Mars. The distances of the centre-planes of Venus and Jupiter are similar while that of the Earth comfortably bounds the Solar System, noting that Pluto reaches a distance of 49.28 AU at aphelion, and reaches inwards to 29.37 AU at perihelion. There is thus a curious progression of 29.37, 39.32, 49.28 and 59.68 for Pluto and the Earth.

If planetary spheres are instead centred on the Earth then they vary greatly in size, and in terms of counterspace packing where the planetary spheres are inside one another we find that interesting inversions take place for Mercury, Venus and Mars. Sometimes Mars is closer to the Earth than Venus in which case its geocentric 'sphere' is smaller than that of Venus, and thus the Venus sphere is *inside* the Mars sphere for counterspace; at other times Mars is *inside* Venus. 'Conjunctions' occur when the spheres are the same size. Are they significant? They are only possible for two planets if the least distance of the outer of the two planets from the Earth is less than the greatest such distance of the inner one. Thus such 'conjunctions' only occur for Mars, Venus and Mercury.

This raises the question of the significance of a region for two counterspace spheres which is *inside* one and *outside* the other, that is, an annular region.

If we allow the idea that human consciousness has evolved since the time of the Greeks then when they referred to planetary spheres, that may have arisen from a consciousness more in tune with the ether of counterspace than is the case for modern people. We have seen that spherical surfaces are related to timeless processes, which also accords with their opinion that beyond the sphere of the Moon all is changeless. They may not have meant by that what we do now, however, that is, that no processes occur out there. The great Danish astronomer Tycho de Brahe triumphantly found change in the heavens when he observed a supernova, showing that he regarded their view of changelessness in the modern way. Yet we have postulated that in counterspace processes may occur that do not reflect into space as requiring time, which brings us up against a fundamental assumption of our modern consciousness: that change always requires time. While for common sense this certainly appears to be true, it is not so simple in quantum physics where some changes are supposed to be timeless, for instance, when an electron jumps from one energy level to another. Worse still some particles are treated as travelling backwards in time, so we cannot really be too complacent with our common sense view. As long as we remain 'onlookers' our common sense works, but as soon as we venture into a participatory consciousness such as the ancients had it starts to break down. An essential feature of quantum physics is that the observer is not merely an 'onlooker' but has to be reckoned as part of the process when making an observation, that is, such delicate observations force us into a participatory role. This points to the future when the clarity and sobriety of our dispassionate 'onlooker' mode can be combined with participation. It is just the chemical ether, closely associated with quantum phenomena, that brings us up against this.

Steiner also referred to lemniscatory paths for the planets, and a shared spiral movement of the Sun and Earth. We have not followed that up here, although it is possible to describe a lemniscatory path of the Earth and Sun that 'saves the appearances' provided a radical re-think of space and light in terms of projective geometry is allowed (Thomas 1975). We hope to integrate these findings into the present work when opportunity and inspiration permit.

Relation between the Earth and the cosmos

The discovery that the standard spherical wave equation may be derived for a linkage between space and counterspace (see Chapter 8) opens up the possibility of interpreting the planetary spheres as hybrid resonant cavities between space and counterspace, enabling the influences of those spheres on the Earth to be put on a scientific rather than a mystical basis. An organ pipe is an example of a resonant cavity which amplifies a particular frequency, and the term 'hybrid' refers to the fact that we have found the possibility of a cavity resonating between space and counter-space. Direct evidence for this has been obtained by Lawrence Edwards (Edwards 1993, Chap. 15) when he found that rhythmic processes occur throughout the winter in supposedly dormant leaf buds. These rhythms relate to cosmic rhythms between the positions of the Moon and the various planets. His work has been dismissed academically despite its high standard of rigour, and despite the fact that it produces 'hard' quantitative evidence. What is embarrassing for the academics is that such things do not fit their paradigm. We suggest that a new paradigm based on the interaction of space and counterspace could allow such findings to be given their due recognition. Indeed when Edwards gave a lecture to professional biologists, he told us that one of them said his work could be so helpful, but that it 'went against their "religion".' We will now describe Edward's discovery in more detail to appreciate its relevance here.

Edwards started a certain line of investigation in 1983 which led to something quite unexpected (Edwards 1993, Chap. 15). He measured the form of leaf buds of trees during the winter, which means that he found the value of a parameter which determines the actual shape of the mathematical egg form describing such buds. Egg forms can be more or less 'pointed,' varying between the extremes of symmetrical prolate spheroids on the one hand and cones on the other. The profiles thus vary between ellipses and triangles. The parameter, designated by λ, is a sensitive measure of where in this spectrum of form the bud lies, noting that it only concerns the shape of the profile and not the spiralling. λ equals 1 for ellipses which are symmetrical, about 1.6 for hens' eggs which are slightly rounder at one end than the other, just above 3 for rose buds which are more distinctly pointed at one end, and infinity for a cone with a triangular cross-section. The leaf buds had their λ value between 2 and 3. Edwards expected the leaf bud shapes to be static through the winter but to his surprise λ varied rhythmically up and down with a roughly two week cycle. As he looked into the matter further he found that different trees had slightly but distinctly different cycle lengths, and these correlated with the times between conjunctions and oppositions of the Moon and a planet, so that for an oak, for example, this correlated closely with the conjunctions and oppositions with Mars. Over nineteen years Edwards collected a large volume of data on this and found a method of analysis which confirms the relationship of different trees to different planets. Interestingly the effect was sharply reduced for a tree near a large electrical transformer. Now it is scarcely possible in reasonable time to grow trees in special locations to test an idea! So he compared the behaviour of knapweed buds under power cables with others and confirmed this result. Recalling what was said earlier about the importance of rhythm we suspect that it is the frequency of the electrical or magnetic oscillation that interferes with a planar linkage between space and counterspace in the dynamic processes involved.

A further confirmation of the planetary influence came when the comet Shoemaker-Levy collided with Jupiter from July 16–22, 1994. Work with knapweed since 1985 had shown it to be connected with Jupiter, but it behaved quite uncharacteristically in 1994 (Edwards, *The Vortex of Life Supplement and Sequel, Vol. 3*). Usually the knapweed λ increases from about 1.3 at the end of June to slightly below 1.7 in mid August, and the results for hundreds of buds analysed between 1985 and 1993 were plotted to give a mean graph of the variation of λ in July and August, together with two graphs of the extreme values of λ reached by individual buds either side of that mean. In 1994 the whole graph lay well above the upper extreme with λ varying from just above 1.7 initially to over 1.8 and then gradually falling again after the impact, but still remaining near 1.8. Truly a remarkable result! (See Figure 30.)

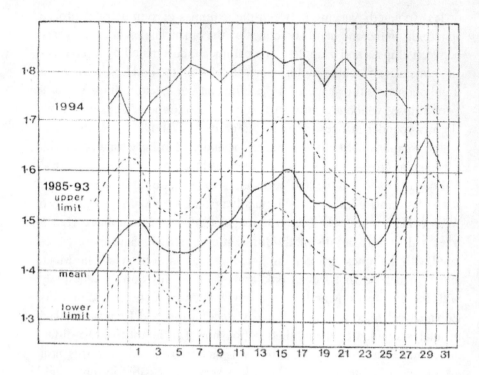

Figure 30. Knapweed graph.

In 1983 the correlation between the rhythmic variation in bud shape and the conjunctions and oppositions of the Moon and planets seemed straight forward, but as the years passed there was a gradual 'slippage' so that although the period of the variations remained correlated the phase gradually changed so that several days between the maxima and minima of the graph and the conjunctions and oppositions began to accrue. After seven years the two were in step again and then the slippage continued as before. It has remained a mystery why this should happen and what governs the rate of slippage. When the possibility of etheric cavities linked to space became apparent it seemed worth investigating whether there could be any connection. Now the size of a cavity varies in inverse proportion to the frequency involved, so that the lowest notes of an organ require the largest pipes and *vice versa*. A frequency of one cycle every two weeks is very low indeed yet the buds as cavities (if that they are) are very small. The scaling between space and counterspace affects the frequency and when the scaling required to relate a bud-sized cavity to a two weekly rhythm was calculated it was very close to the reciprocal of the so-called velocity of light (which for us is not a velocity but a scaling constant). It is nearly exact for a 4 mm bud, and many buds are this sort of size, including leaf buds, but of course there is large variation more generally. This seemed suggestive and a parallel investigation of the scaling of space and counterspace showed that for some linkages the scaling constant is indeed the reciprocal of that for others, and in particular it appears this is true for the scaling of the life ether. We do not expect the scaling constant to be the only factor involved, and neither do we expect it always to be the reciprocal of the scaling for light, for the substances themselves can affect scaling as we saw in the case of refraction. So variation in bud size is not an instant show-stopper. But the scaling constant for light is so large and thus its reciprocal so small that to arrive immediately within the right 'ball park' seems interesting indeed. What it suggests is quite another kind of cavity based not on a heat/warmth cycle but on a form/life-ether cycle.

The connection between the cosmos and the bud then suggests a real connection between rhythmic processes on Earth and in the cosmos. Thus it makes sense to suggest that further exploration of such relationships should be based on rhythmic processes, of which there are many. It also indicates that life processes generally may relate to cosmic rhythms, which could have wide practical implications, the discovery of the interference of power lines being an example. So far only one case has been explored with any rigour in the present context, but that one important case makes all the difference between merely speculating that such rhythmic connections exist and knowing that such an idea already has experimental backing. It also focusses such enquiries on to rhythms rather than, say, relative positions of cosmic bodies.

12. Phenomena

In this chapter we will attempt to indicate how counterspace ideas may be applied to the understanding of some selected phenomena.

Planar linkages

We need to understand planar linkages for some of these phenomena. By planar linkages we mean that planes may belong to both spaces at once. We need to distinguish between an extensive linkage in which the whole plane is linked, and an intensive one. An extensive linkage could be very dangerous as any CSI in the plane which is of the same type as that for which the linkage originated would become involved, possibly leading to damage and destruction. For the extent of such a linkage would be infinite, so a very small rotation of the plane would result in very large disturbances at a distance from the source.

An intensive linkage in contrast would be such that there is a local planar linkage occurring at a point which does not extend in ordinary space but does define a planar orientation at that point. In counterspace the plane would be extensive, of course, because it is a fundamental element of that space. To illustrate this, imagine a green plane containing a red point. Then the counterspace aspect of the linkage is illustrated by the green plane, and the spatial aspect by the planar orientation at the red point. A bivector (explained in Chapter 7) would be the likely linkage element as we are concerned with a plane. This is more process-oriented and less abstract than the idea of an extensive linkage. We saw earlier on that intensive linkages operate via rates of change of quantities (see Chapter 4). Thus such a linkage may be expected to arise where planar orientations are changed by a process. Thus we do not envisage permanent static planar linkages, but rather that such linkages arise while a process of some kind is occurring; we will see some examples below.

However an extensive effect of such intensive linkages, of a nondestructive kind, could nevertheless manifest via the primal counterspace of the CSIs involved. The difference is that the distant effect would be local there, unlike the very large disturbance that a purely extensive linkage would cause. For example a rotation of the linked plane would result in a similar rotation (or perhaps tendency to rotate) in a plane at the distant point. In other words another local planar linkage would arise there imaging the first.

Strain and stress arise if linked planes for different CSIs have different turns in from infinity, that is, spatially speaking they are at different distances from their parent CSIs. This operates only in counterspace where the plane is extensive and relates to other CSIs apart from the linkage source.

A further aspect concerns the CSI of the primal counterspace (see Chapter 4). A linked plane must of necessity be related to that primal counterspace infinity as it is the source of all its 'daughter' CSIs, and the orientation of the counterspace aspect of the plane may differ from that at the source CSI.

We are not ruling out the possibility of extensively linked planes, but we do not normally expect to encounter them as their destructive properties are not observed in nature.

Water vortices

A water vortex appears to be 'woven' of planar linkages as the rotation brings about surfaces through which the water flows, giving changing planes for which the planar linkages envisaged would be tangential to these surfaces. In the surface of the vortex these planes are at various angles at different positions in the surface, suffering mutual turn stress, which is relieved as the vortex flattens out, that is, the rotation slows down. If the rotation is forcibly stopped by counter-stirring then some of the stress in counterspace may be trapped because a smooth transition between spatial energy and counterspatial stress is prevented. Thus the relation between the spaces established by the rotation is as-it-were 'caught' in the water, retaining a cosmic connection as a result. This may also be understood by noting that a water vortex is related to path curves (see below) with a connection between the cosmic periphery and a CSI. If the vortex 'runs down' that connection is gracefully relinquished, but if it is suddenly halted, especially in a way that leads to the formation of many drops of water, then the cosmic connection may be transferred to those drops, and retained. This effect would be enhanced for any substances dissolved in the water where more than one primal counterspace would be involved.

Now biodynamic preparations for farming are prepared precisely by this sort of action, so we may see how etheric forces are caught into those preparations. Also flowforms invented by John Wilkes (see Wilkes 2003 for details) introduce counter-rotation and a region of chaos which may be expected to have the same result. Flowforms have been used successfully for making biodynamic preparations, as well as increasing the rate of seed germination by up to 11% (*op. cit.* p. 141). Flowforms also exhibit rhythmic behaviour which can be a bridge between space and counterspace. It seems that the tone ether links to rhythms very readily; recall trying to be quiet at night, when every tiny movement seems to make a huge noise! The tone ether gives the tone or sound that we actually experience which is incorporated in the spatial sound waves. Thus where rhythms arise we may expect its presence, and particularly in hybrid resonant cavities (see Chapter 11) it is envisaged to be a way of 'priming' them. In our investigations, it was observed that the frequencies of many flowforms were close to one that can be associated with a resonant cavity related to the Sun.

Water vortices also played a central role in the work of Viktor Schauberger (Alexandersson 1982 and Coates 1996). He claimed to have access to what we have referred to as etheric forces, and some of his observations and experiments seemed to bear that out. Again we could interpret his findings as being concerned with planar linkages between space and counterspace of the intensive variety in most cases. However the large forces he claimed to have accessed in his 'trout turbine' and 'flying saucer' are the one source of evidence, if valid, for the possible existence of extensive planar linkages. It would seem the 'implosive' technology he sought so valiantly to introduce would produce physical effects via linkages between space and counterspace, thus engendering those effects by setting up a strain in counterspace that reacted via the linkage on spatial objects. This would be the general way in which so-called 'etheric technology' might work.

Water vortices produced in closed containers, that is, where the water does not flow out of the base, show an interesting sensitivity at their cusp. This cusp, at the bottom of the vortex, often moves erratically as if 'avoiding' something. This led to the suspicion that the planar linkages are avoiding the centre of the Earth. If, as we suspect, there is a counterspace infinity at the centre of the Earth which is that of a 'primal counterspace,' then should a planar linkage contain that centre it would have 'gone to infinity' in counterspace and hence would become 'stuck' there. However that would also require the plane to have executed an infinite turn. Hence the planar linkages at the cusp of a vortex appear to be avoiding such an alignment. Elsewhere in the surface the tangent planes are generally in no danger of containing the Earth's centre. A more detailed consideration taking into account the ripples in the surface increases the likelihood of vertical planes down near the cusp, enhancing the wriggling that is observed.

Crystal growth

Crystals are a beautiful product of the mineral kingdom which have been studied extensively over the years. They may have a number of distinct forms such as the cube, dodecahedron, hexagonal prism and so on which are easily described mathematically. Three striking phenomena are that they grow, that their surfaces tend to be flat, and that the angles between their faces are precisely formed in the ideal case. Conventionally they are explained with reference to atomic bonds where the angles at which bonds between atoms occur are very precisely defined, so three-dimensional lattices arise with the component atoms at the vertices. A simple case is the cubic lattice composed of lines mutually at right angles and with a fixed minimum spacing between those lines, all other spacings being simple multiples of that minimum. The lattice thus looks like an array of stacked cubes.

We may also expect lattices to arise through the minimization of planar linkage stress arising in counterspace. A regular array of CSIs in this way has minimum stress compared with a disordered collection of CSIs. As we remarked earlier we only expect planar linkages to arise when some process is taking place, and in this case that is the process of crystallization. The growth is then controlled by the tendency to minimize the stress, and so the regular form arises as observed. No planar linkages are expected to remain in the finished crystal when the process has stopped. The accurately flat surfaces observed in crystals seem more easily understood holistically in terms of planes. How accurate, in contrast, must the conventionally conceived bond angles be to achieve such flatness, for instance, where of the order of a billion such bonds are involved across 1 cm.

It is also found that 60 and 120 degree angles between crystal surfaces are particularly stable with respect to the cosmic and inner infinities of the two spaces, as the turn in from infinity of each face may be shown to equal that between two such faces. Apart from 90 degree angles, other angles are not stable, leading either to 'frozen stress' or else to unstable spiral growth or decay which does not generate a stable overall form. Angles of 90 degree are also stable but for a different reason. Now it is well known in crystallography that the only permitted symmetry angles are 360, 180, 120, 90 and 60 degrees.

We saw in Chapter 9 that the life ether operates through fully metric linkages. Now crystals have a distinctly metric appearance with their regular lattice-like structure, and are quintessentially mineral in nature. However while the process of crystallization is occurring, with the attendant planar linkages, that process may be influenced by any life ether forces present via those planar linkages. So we see the wonderful organic-style forms arising in frost patterns, and in the research done on sensitive crystallization as a method of rendering etheric forces visible (cf. Pfeiffer 1936). The interaction between space and counterspace here leads to the observed forms in conjunction with the physical factors investigated by Howard Smith (Smith 1975). Indeed without those physical constraints no forms would be visible. But they alone do not fully explain the forms. Concentration depletion is perhaps the most important physical factor, for as crystals grow the concentration of the solution in their immediate neighbourhood is decreased, limiting that growth and impeding other immediately adjacent growth. Hence we obtain many small often fractal-like forms with intervening space or reduced thickness so that the forms stand out by contrast. They are often quite organic in appearance. We would expect a global coherence of form across much or all of the test dish if indeed the planar linkages are affected by a life ether 'field,' but not otherwise. That is the basis of interpreting this test method.

Crystals, being solid and in some cases elastic, can be made to vibrate. Harry Oldfield has been conducting some fascinating research into methods of picturing the human aura (see Jane and Grant Solomon's account). He uses electrical fields to 'probe' the aura, and has devised a method to scan it and produce a picture on a computer screen. His aim has been to treat sickness by stimulation using crystals, called Electro-Crystal Therapy, apparently with considerable success. At first sight one might object that as the aura is non-physical this must be flawed, but on the basis of linkages between space and counterspace it can perhaps be understood differently. Living tissue is intimately linked to the etheric via counterspace, and appropriate physical stimulation of it may set up a stress in counterspace which reacts back on the physical in a measurable way. Thus the pictures may not be directly of the aura, but the images may certainly bear some indirect correlation with it. Likewise stimulation of the physical body using the right frequency set up in crystals may react back on the etheric body in a healing manner. For then a process is initiated which can rekindle the planar linkages.

Path curves

Lawrence Edwards made extensive investigations of organic forms in relation to geometry and counterspace (Edwards 1982 and 1993). At the end of the nineteenth century, the mathematician Felix Klein discovered special curves that in the literature in English are called *path curves*. We have encountered the idea of geometric transformation, and a simple example is rotation where we imagine the whole of space turning about an axis. This can be described mathematically as a process called a transformation which moves every point, line and plane in space round the axis. Thus a point one metre from the axis will be moved round a circle of one metre radius lying in a plane at right angles to the axis. All the points of space except those on the axis are set in motion in this way. The circles round which points move, which includes all possible circles centred on the axis lying in planes at right angles to it, are the path curves of this transformation. What is special about the circles is that each one is, as a whole, left unchanged by the transformation: it does not change radius or position. Many kinds of transformation are possible, but we restrict ourselves to those that preserve straightness and flatness. This means that when a straight line is moved to another position it ends up straight after the move, and likewise planes remain flat after they have been moved. This is illustrated for a simple case in Figure 31.

Three points A, B, C on a straight line in a (horizontal) plane (α) are projected from a point P on to another (vertical) plane (β) as A_1 B_1 C_1. The lines PA, PB, PC lie in a plane which meets plane β in a straight line, so the points A_1 B_1 C_1 lie on that straight line. Any projective transformation is composed of a series of transformations such as above, so straightness is preserved throughout.

Another example is expansion about a central point where all other points move outwards from the centre along lines through it. Here the

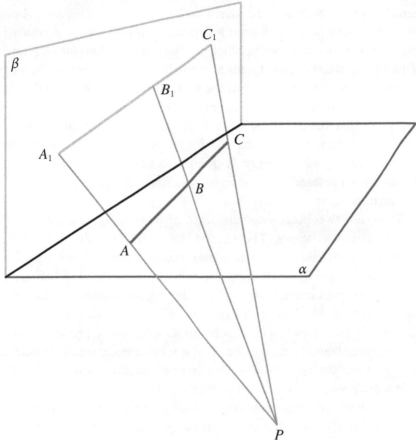

Figure 31. A projective transformation.

path curves are those lines. If we combine rotation and expansion then we get spiral path curves. George Adams became interested in a particular kind of path curve that spirals round an egg shaped surface on the one hand, or round a vortex shaped surface on the other (Adams and Whicher 1980, p. 213). He pointed out that the curves on egg shapes look very like the spirals seen on pine cones and plant buds, and he saw these curves as arising from an interplay between space and counter-space. Edwards investigated how well such curves fit the actual forms found in nature, and the results were strikingly good (Edwards 1993, Chap. 5). It has been objected that this is 'mere curve fitting,' but that entirely misses the point. First of all the kinds of spiral are very narrowly defined, and yet fit closely a wide variety of natural forms such as pine cones, many kinds of plant bud, birds eggs of many varieties, the left ventricle of the heart and the uterus during pregnancy. The vortical ones fit water vortex profiles closely. It is possible to find some mathematical curve to fit almost anything, which is of course not of scientific interest, but here we are concerned with a narrowly and precisely specified type of curve which alone is used in these various cases. The second reason for the research lies not in an attempt to 'find something that fits' but to apply the idea that counterspace is a real factor in nature and to test that, in other words, the work is based on a thought context. In conventional science some kind of theory is thought to be mandatory as a context within which research is conducted, which we interpret more widely as a thought context.

There are two polar opposite ways of describing a path curve, the simpler being as above. The second is to start with planes instead of points in which case the transformation moves a plane in such a way that it 'moulds' a curve, which turns out to be the same kind of path curve. Adams saw path curves especially in this second sense and thus imagined counterspace moulding physical form, so that both the plane-based and point-based views operate together as a formative process weaving between space and counterspace. This is what is being tested, in contrast to mere form fitting. Edwards' discovery of cosmic rhythms in relation to leaf buds was described in Chapter 11.

Path curves illustrate how planar linkages arising during the process of growth are local at the points of the path curve where those planes touch it. A detailed exploration of how path curves arise through reciprocal transformations between space and counterspace (Thomas 1984) also shows the role of the polarity between point and plane in path curves.

Solutions

When a substance (the solute) dissolves in a liquid (the solvent) it disappears from sight and loses its solid form, for instance salt dissolving in water. The latter pervades the whole of the solvent when dissolved, rather like a gas. This suggests that the solute, if solid, starts out with a metric linkage between space and counterspace, which becomes affine when it dissolves, as affine linkages characterize gases. The CSIs of the solute will suffer polar volume strain in the way we saw when considering the ideal gas law, and thus we would expect them to be redistributed to minimize that, namely throughout the solvent. We note that the process of dissolving is accelerated by stirring, suggesting that the dynamic process involved possibly promotes planar linkages as well. This linkage type-change may indicate that a linkage is established between the volume of the solvent and the polar volume of the dissolved solute, the latter being of a planar character.

We may view homeopathy in the light of these observations. The chemical ether as tone ether manifests in gases when waves are present, and we may thus postulate that it can manifest in the 'solute-gas' in the solution. This is not necessarily the case, however, and for homeopathy we see why the procedure may or may not work, as it depends upon whether oscillations have been engendered during the dilution process. It has also been observed that some experimenters inhibit such processes while others enable it (McTaggart 2003, p. 95). Thus succussion, the process used to dilute the original solution by stages, must be such that waves are induced in the solute-gas, ushering in the tone ether. Succussion is the procedure where for instance, a solution is diluted by adding distilled water, and then the container is shaken or 'banged' forty or more times before diluting again. Samuel Hahnemann, the inventor of homeopathy, is said to have banged the bottle on the family Bible!

Furthermore, the strength of the waves may depend upon the concentration, giving rise to the observed rhythmic relation between efficacy and potency (concentration), where a graph of efficacy against potency rises and falls rhythmically. The efficacy then depends upon the waves rather than the solute as substance. Thus we have a kind of 'music' in a potentized substance, which can persist in physically pure distilled water as it does not depend upon any physical solute being present.

Another way of viewing the succussion is similar to that explained above in connection with vortices: if the liquid is caused to break into drops by succussion then the music may be transferred into the polar volume of those drops, and retained after they merge again. The whole point is to establish a connection between the etheric aspect of the solute in its polar volume (once it is dissolved) and the mineral aspect of the solvent in its ordinary volume. This may suggest other more efficient methods of preparing potentized remedies, by introducing the music directly rather than by repeated dilution. That could be developed further when an apparatus to sense such waves has been developed, which would probably exploit the hybrid resonant cavity effect mentioned before.

13. Looking Forward

This book has described work in progress rather than a finished result. The aim has been to throw light on Rudolf Steiner's research which may appear strange without a proper context, and to take further the work on counterspace done by George Adams. It represents only a beginning with much work to be done. Developing a new paradigm is after all an ambitious project, competing as it does with the detailed work of so many Nobel laureates. Doubtless the work of many unmentioned researchers could have been fruitfully included; my apologies to them all.

Another aim of the work is to place science into a wider more spiritual context. This work begins from where we are, in the physical world with a point-centred consciousness, and explores what happens if another aspect of the world accessible to a 'peripheral consciousness' is taken into account. The linkages between the two spaces bring in a third aspect related to the *will* of beings, for linkages are not in reality abstract. The term 'spiritual' is not intended in a religious sense but as relating to that other holistic aspect of the world accessible to those able to change their consciousness (temporarily) into a peripheral one. It is best approached philosophically as described by Rudolf Steiner in his *Philosophy of Freedom* where 'thinking about thinking' leads to a direct and lucid awakening to a spiritual activity: your own thinking activity. The role of that and of will is then seen to be the basis of linkages. This leads to a scientific rather than a religious approach to the spiritual aspects of the world. It is found that the thinking activity of human beings is deeply connected with the ethers, especially the light ether. The work presented here can, it is hoped, be a bridge across the threshold lying between our ordinary awareness and the meticulous science that goes with it, and a widened more comprehensive science — no less meticulous — which does justice also to the cosmic peripherally-based aspects of the one reality which is our shared world.

The further outlook envisaged is to derive practical instruments based on the principles described, to understand matter and ether more clearly, and to find solutions to current practical problems such as the treatment of nuclear waste. Above all, to understand ever better the relation of the manifest world to its unmanifest subtle aspects which are everywhere indicated. The hope is that many who for good reasons stick to the materialistic paradigm may come to see that more is possible than that, enabling many urgent practical issues to be approached in a broader and more genuinely holistic manner. Thus the possibility of quantitative results via counterspace strain even for subtle realms may provide a suitable 'bridge.'

Counterspace is by nature non-local, spanning all CSIs. It enables instantaneous interactions between them and is the basis for the action of fields. Where Lynne McTaggart in her fascinating book *The Field* sought a physical basis for everything, we see counterspace as the basis sought for the more subtle aspects of reality uncovered by the researchers she describes. The 'zero-point field' of physics she embraces can be seen as the interface between two spaces rather than a mysterious exploitation of the uncertainty principle. Indeed the latter can be explained by the mismatch between affine and metric space (Thomas 1999, p. 130) which must be resolved when a measurement is made. The evidence for a cosmic unity and the non-material interactions of cells in biology she describes, as well as the kind of phenomena found in homeopathy, are — we suggest — mediated by counterspace. The challenge remains to flesh this out, to understand better the intrinsic nature of the ethers in their own terms rather than in material terms. The present author's ongoing research concerns the use of the waves found in relation to hybrid cavities to provide the basis for what otherwise is ascribed to the zero-point field, as well as to re-describing the structure of matter. For such waves can be very rich in structure.

To end on a philosophical note, the broader results of modern physics, engineering, biology and indeed all of natural science rest upon quantum physics. The latter does not explain anything, in the sense that it has no content apart from mathematics. It predicts the probability of results with amazing accuracy, but since a conference in Copenhagen in 1931 physicists have declined to say *how* they come about. The content that is missing must be sought elsewhere. Thus no explanation in the usual sense is forthcoming from science, and it has no basis to pontificate on ultimate questions lying outside its field of competence, such as whether there are spiritual forces at work in our world. Natural science abstracts its results from sense experience, discarding the latter in the process (cf. Chapter 1 concerning the role of qualia), but what is suggested in this book is that other realms of experience exist, one of which has been explored here, that may lead towards an approach to those questions lying outside the remit of natural science.

Bibliography

Adams Kaufmann, George (1933) *Space and the Light of Creation*, self-published, London.

Adams, George (1979) The *Lemniscatory Ruled Surface in Space and Counterspace*, Rudolf Steiner Press, London.

—, (1977) *Universal Forces in Mechanics*, Rudolf Steiner Press, London.

—, and Whicher, Olive (1980) *The Plant Between Sun and Earth*, Rudolf Steiner Press, London.

Ayer, A.J. (1970) *Language, Truth & Logic*, Victor Gollancz, London.

Collins, John (1997) *Perpetual Motion, an Ancient Mystery Solved?* Permo Publications, Leamington Spa.

Edwards, Lawrence (1982) *The Field of Form*, Floris Books, Edinburgh.

—, (1993) *The Vortex of Life*, Floris Books, Edinburgh.

—, *The Vortex of Life Supplement and Sequel*, Vol. 3, published privately, available on www.vortexoflife.org.uk

French A.P. and Taylor E.F. (1979) *An Introduction to Quantum Physics*, Van Norstrand Reinhold.

Gleick, James (1987) *Chaos*, Sphere Books.

Heisenberg, Werner (1949) The *Physical Principles of the Quantum Theory*, Dover Publications, first published in 1930 by Chicago Press.

—, (1979) *Philosophical Problems of Quantum Physics*, Ox Bow Press, Woodbridge, Connecticut, first published in 1952 by Pantheon.

Jaynes, Julian (1990) *The Origin of Consciousness in the Breakdown of the Bicameral Mind*, Penguin Books, London.

McTaggart, Lynne (2003) *The Field*, Element Books, UK.

Penrose, Roger (1994) *Shadows of the Mind*, Oxford University Press.

Rudnicki, Konrad (1995) *The Cosmological Principles*, Jagiellonian University, Krakow.

Solomon, Jane and Grant (1998) *Harry Oldfield's Invisible Universe*, Thorsons.

Steiner, Rudolf (1920) *Heat Course,* March 1920, Stuttgart (GA 321).

—, (1986) *True and False Paths of Spiritual Investigation*, August 1924. Tr. A. Parker, Rudolf Steiner Press, London (GA 243).

—, (1920) *Man Hieroglyph of the Universe*, Dornach 1920.

—, (1992) *The Philosophy of Freedom, a Philosophy of Spiritual Activity*. Tr. Rev. R. Stebbing, Rudolf Steiner Press London (GA 4).

Thomas, N.C. (2008) *Science Between Space and Counterspace,* Temple Lodge Press, London.

—, (1984) 'Das Zusammenwirken vom Raum und Gegenraum zur Erzeugen von Wegkurven,' in *Mathematisch-Physikalische Korrespondenz,*

No 133, July 1984, Mathematical-Astronomical Section, Dornach, Switzerland.

—, (1975) 'An Approach to the Lemniscate Path of Sun and Earth,' *Mathematical-Physical Correspondence,* No 13, Michaelmas 1975.

—, web page at *www.nct.anth.org.uk*

Velmans, Max (2000) *Understanding Consciousness,* Routledge, London.

Wilkes, John (2003) *Flowforms, the Rhythmic Power of Water,* Floris Books, Edinburgh.

Index

Floris Books

For news on all our **latest books**,
and to receive **exclusive discounts**,
join our mailing list at:

florisbooks.co.uk

Plus subscribers get a FREE book
with every online order!

We will never pass your details to anyone else.

Printed in the USA
CPSIA information can be obtained
at www.ICGtesting.com
JSHW051457171023
50342JS00009B/283